Thomas Sterry Hunt

The Geognosy of the Appalachians and the Origin of Crystalline Rocks

Address to the American Association for the Advancement of Science

Thomas Sterry Hunt

The Geognosy of the Appalachians and the Origin of Crystalline Rocks
Address to the American Association for the Advancement of Science

ISBN/EAN: 9783337059699

Printed in Europe, USA, Canada, Australia, Japan

Cover: Foto ©berggeist007 / pixelio.de

More available books at **www.hansebooks.com**

ADDRESS

TO THE

AMERICAN ASSOCIATION

FOR THE

ADVANCEMENT OF SCIENCE,

BY

ERRATUM.

For foot notes on page 45, read

* Pogg. Annal. lxviii, 319.
† Amer. Jour. Sci., II, xvi, 218.

SALEM:

NATURALISTS' AGENCY.

1871.

ADDRESS

TO THE

AMERICAN ASSOCIATION

FOR THE

ADVANCEMENT OF SCIENCE,

BY

THOMAS STERRY HUNT, LL.D.

ON RETIRING FROM THE OFFICE OF PRESIDENT OF THE ASSOCIATION.
INDIANAPOLIS, AUGUST 16, 1871.

Printed in advance from the Association Number of the American Naturalist.

SALEM:

NATURALISTS' AGENCY.

1871.

ADDRESS

OF

THOMAS STERRY HUNT,

ON RETIRING FROM

The office of President of the American Association for the Advancement of Science.

[Delivered at the 20th meeting of the Association, at Indianapolis, August 16th, 1871.]

GENTLEMEN OF THE AMERICAN ASSOCIATION FOR THE ADVANCE-
MENT OF SCIENCE : —

IN coming before you this evening my first duty is to announce
the death of Professor William Chauvenet. This sad event was
not unexpected, since, at the time of his election to the presidency
of the Association, at the close of our meeting at Salem in August,
1869, it was already feared that failing health would prevent him
from meeting with us at Troy, in 1870. This, as you are aware,
was the case, and I was therefore called to preside over the Association in his stead. In the autumn of 1869, he was compelled by
illness to resign his position of Chancellor of the Washington University of St. Louis, and in December last died at the age of fifty
years, leaving behind him a record to which science and his country
may point with just pride. During his connection of fourteen years
with the Naval Academy at Annapolis he was the chief instrument
in building up that institution, which he left in 1859 to take the
chair of Astronomy and Mathematics at St. Louis, where his remarkable qualities led to his selection, in 1862, for the post of

(3)

chancellor of the university, which he filled with great credit and
usefulness up to the time of his resignation. * It is not for me to
pronounce the eulogy of Professor Chauvenet, to speak of his profound attainments in astronomy and mathematics, or of his published works, which have already taken rank as classics in the
literature of these sciences. Others more familiar with his field
of labor may in proper time and place attempt the task. All who
knew him can however join with me in testifying to his excellencies
as a man, an instructor and a friend. In his assiduous devotion
to scientific studies he did not neglect the more elegant arts, but
was a skilful musician, and possessed of great general culture and
refinement of taste. In his social and moral relations he was
marked by rare elevation and purity of character, and has left to
the world a standard of excellence in every relation of life which
few can hope to attain.

In accordance with our custom it becomes my duty in quitting
the honorable position of president, which I have filled for the
past year, to address you upon some theme which shall be ger-
mane to the objects of the Association. The presiding officer, as
you are aware, is generally chosen to represent alternately one of
the two great sections into which the members of the Association
are supposed to be divided ; viz., the students of the natural-history sciences on the one hand, and of the physico-mathematical
and chemical sciences on the other. The arrangement by which,
in our organization, geology is classed with the natural-history
division, is based upon what may fairly be challenged as a somewhat narrow conception of its scope and aims. While theoretical
geology investigates the astronomical, physical, chemical and biological laws which have presided over the development of our
earth, and while practical geology or geognosy studies its natural
history, as exhibited in its physical structure, its mineralogy and
its paleontology, it will be seen that this comprehensive science is
a stranger to none of the studies which are included in the plan of
our Association, but rather sits like a sovereign, commanding in
turn the services of all.

As a student of geology, I scarcely know with which section of
the Association I should to-day identify myself. Let me endeavor

---

* Amer. Jour. Sci., III, i, 233.

rather to mediate between the two, and show you somewhat of the two-fold aspect which geological science presents, when viewed respectively from the stand-points of natural history and of chemistry. I can hardly do this better than in the discussion of a subject which for the last generation has afforded some of the most fascinating and perplexing problems for our geological students; viz., the history of the great Appalachian mountain chain. Nowhere else in the world has a mountain system of such geographical extent and such geological complexity been studied by such a number of zealous and learned investigators, and no other, it may be confidently asserted, has furnished such vast and important results to geological science. The laws of mountain structure, as revealed in the Appalachians by the labors of the brothers Henry D. and William B. Rogers, of Lesley and of Hall, have given to the world the basis of a correct system of orographic geology,* and many of the obscure geological problems of Europe become plain when read in the light of our American experience. To discuss even in the most summary manner all of the questions which the theme suggests, would be a task too long for the present occasion, but I shall endeavor to-night in the first place to bring before you certain facts in the history of the physical structure, the mineralogy and the paleontology of the Appalachians; and in the second place to discuss some of the physical, chemical and biological conditions which have presided over the formation of the ancient crystalline rocks that make up so large a portion of our great eastern mountain system.

I. The Geognosy of the Appalachian System.

The age and geological relations of the crystalline stratified rocks of eastern North America have for a long time occupied the attention of geologists. A section across northern New York, from Ogdensburg on the St. Lawrence to Portland in Maine, shows the existence of three distinct regions of unlike crystalline schists. These are the Adirondacks to the west of Lake Champlain, the Green Mountains of Vermont, and the White Mountains of New Hampshire. The lithological and mineralogical differences between the rocks of these three regions are such as to have attracted the attention of some of the earlier observers. Eaton, one of the

founders of American geology, at least as early as 1832, distinguished in his Geological Text book (2d edition) between the gneiss
of the Adirondacks and that of the Green Mountains. Adopting
the then received divisions of primary, transition, secondary and
tertiary rocks, he divided each of these series into three classes,
which he named carboniferous, quartzose and calcareous; meaning
by the first schistose or argillaceous strata such as, according to
him, might include carbonaceous matter. These three divisions in
fact corresponded to clay, sand, and lime-rocks, and were supposed
by him to be repeated in the same order in each series. This was
apparently the first recognition of that law of cycles in sedimentation upon which I afterwards insisted in 1863.* Without, so far as I
am aware, defining the relations of the Adirondacks, he referred to
the lowest or carboniferous division of the primary series, the crystalline schists of the Green Mountains, while the quartzites and
marbles at their western base were made the quartzose and calcareous divisions of this primary series. The argillites and sandstones
lying still farther westward, but to the east of the Hudson River,
were regarded as the first and second divisions of the transition
series, and were followed by its calcareous division, which seems to
have included the limestones of the Trenton group; all of these
rocks being supposed to dip to the westward, and away from the
central axis of the Green Mountains. Eaton does not appear
to have studied the White Mountains, or to have considered their
geological relations. They were, however, clearly distinguished
from the former by C. T. Jackson in 1844, when, in his report on
the geology of New Hampshire, he described the White Mountains as an axis of primary granite, gneiss and mica-schist, overlaid successively, both to the east and west, by what were designated by him Cambrian and Silurian rocks; these names having, since
the time of Eaton's publication, been introduced by English geologists. While these overlying rocks in Maine were unaltered, he
conceived that the corresponding strata in Vermont, on the western
side of the granitic axis, had been changed by the action of intrusive
serpentines and intrusive quartzites, which had altered the Cambrian into the Green Mountain gneiss, and converted a portion of
the fossiliferous Silurian limestones of the Champlain valley into
white marbles.† Jackson did not institute any comparison be-

---

* Amer. Jour. Sci., II, xxxv, 166.
† Geology of New Hampshire, 160-162.

tween the rocks of the White Mountains and those of the Adirondacks; but the Messrs. Rogers in the same year, 1844, published an essay on the geological age of the White Mountains, in which, while endeavoring to show their Upper Silurian age, they speak of them as having been hitherto regarded as consisting exclusively of various modifications of granitic and gneissoid rocks, and as belonging "to the so-called primary periods of geologic time."[*] They however considered that these rocks had rather the aspect of altered paleozoic strata, and suggested that they might be, in part at least, of the age of the Clinton division of the New York system; a view which was supported by the presence of what were at the time regarded by the Messrs. Rogers as organic remains. Subsequently, in 1847,[†] they announced that they no longer considered these to be of organic origin, without however retracting their opinion as to the paleozoic age of the strata. Reserving to another place in my address the discussion of the geological age of the White Mountain rocks, I proceed to notice briefly the distinctive characters of the three groups of crystalline strata just mentioned, which will be shown in the sequel to have an importance in geology beyond the limits of the Appalachians.

I. *The Adirondack or Laurentide Series.* The rocks of this series, to which the name of the Laurentian system has been given, may be described as chiefly firm granitic gneisses, often very coarse-grained, and generally reddish or grayish in color. They are frequently hornblendic, but seldom or never contain much mica, and the mica-schists, (often accompanied with staurolite, garnet, andalusite and cyanite), so characteristic of the White Mountain series, are wanting among the Laurentian rocks. They are also destitute of argillites, which are found in the other two series. The quartzites, and the pyroxenic and hornblendic rocks, associated with great formations of crystalline limestone, with graphite, and immense beds of magnetic iron ore, give a peculiar character to portions of the Laurentian system.

II. *The Green Mountain Series.* The quartzo-feldspathic rocks of this series are to a considerable extent represented by a fine-grained petrosilex or eurite, though they often assume the form of a true gneiss, which is ordinarily more micaceous than the typical

---

* Amer. Jour. Sci., II, i, 411.
† Ibid, II, v, 116.

Laurentian gneiss. The coarse-grained, porphyritic, reddish varieties common to the latter are wanting in the Green Mountains, where the gneiss is generally of pale greenish and grayish hues. Massive stratified diorites, and epidotic and chloritic rocks, often more or less schistose, with steatite, dark colored serpentines and ferriferous dolomites and magnesites also characterize this gneissic series, and are intimately associated with beds of iron ore, generally a slaty hematite, but occasionally magnetite. Chrome, titanium, nickel, copper, antimony and gold are frequently met with in this series. The gneisses often pass into schistose micaceous quartzites, and the argillites, which abound, frequently assume a soft, unctuous character, which has acquired for them the name of talcose or nacreous slates, though analysis shows them not to be magnesian, but to consist essentially of a hydrous micaceous mineral. They are sometimes black and graphitic.

III. *The White Mountain Series.* This series is characterized by the predominance of well defined mica-schists interstratified with micaceous gneisses. These latter are ordinarily light colored from the presence of white feldspar, and, though generally fine in texture, are sometimes coarse-grained and porphyritic. They are less strong and coherent than the gneisses of the Laurentian, and pass, through the predominance of mica, into mica-schists, which are themselves more or less tender and friable, and present every variety, from a coarse gneiss-like aggregate down to a fine-grained schist, which passes into argillite. The micaceous schists of this series are generally much richer in mica than those of the preceding series, and often contain a large proportion of well defined crystalline tables belonging to the species muscovite. The cleavage of these micaceous schists is generally, if not always, coincident with the bedding, but the plates of mica in the coarser-grained varieties are often arranged at various angles to the cleavage and bedding-plane, showing that they were developed after sedimentation, by crystallization in the mass; a circumstance which distinguishes them from rocks derived from the ruins of these, which are met with in more recent series. The White Mountain rocks also include beds of micaceous quartzite. The basic silicates in this series are represented chiefly by dark colored gneisses and schists, in which hornblende takes the place of mica. These pass occasionally into beds of dark hornblende-rock, sometimes holding garnets. Beds of crystalline limestone occasionally occur in the schists of the White Mountain series, and are sometimes accompanied by pyroxene, garnet, idocrase, sphene and graphite, as in the corresponding rocks of the Laurentian, which this series, in its more gneissic portions, closely resembles, though apparently distinct geognostically. The limestones are intimately associated with the highly micaceous schists containing staurolite, andalusite, cyanite and garnet. These schists are sometimes highly plumbaginous, as seen in the graphitic mica-schist holding garnets in Nelson, New Hampshire, and that associated with cyanite in Cornwall, Conn. To this third series of crystalline schists belong the concretionary granitic veins abounding in beryl, tourmaline and lepidolite, and occasionally containing tinstone and columbite. Granitic veins in the Laurentian gneisses frequently contain tourmaline, but have not, so far as yet known, yielded the other mineral species just mentioned. *

Keeping in mind the characteristics of these three series, it will be easy to trace them southward by the aid of the concise and accurate descriptions which Prof. H. D. Rogers has given us of the rocks of Pennsylvania. In his report on the geology of this state he has distinguished three districts of various crystalline schists, which are by him included together under the name of gneissic or hypozoic rocks. Of these districts the most northern, or the South Mountain belt, to the northwest of the Mesozoic basin, is said to be the continuation of the Highlands of New York and New Jersey, which, crossing the Delaware near Easton, is continued southward through Pennsylvania and Maryland into Virginia, where it appears in the Blue Ridge. The gneiss of this district in Pennsylvania is described as differing considerably from that of the southernmost district, being massive and granitoid, often hornblendic, with much magnetic iron, but destitute of any considerable beds of micaceous, talcose or chloritic slate, which mark the rocks of the southern district. These characters are sufficient to show that the gneiss of this northern district is lithologically as well as geognostically identical with that of the Highlands, and belongs like it to the Adirondack or Laurentian system of crystalline rocks. The gneiss of the middle district of Pennsylvania, to the south of the Mesozoic, but north of the Chester valley, is described by Rogers as resembling that of the South Mountain or

* Hunt, Notes on Granitic Rocks; Amer. Jour. Sci., III, i, 182.

northern district, and to consist chiefly of white feldspathic and dark hornblendic gneiss, with very little mica, and with crystalline limestones.

The gneiss of the third or southern district, that lying to the south of the Montgomery and Chester valleys, comes from beneath the Mesozoic of New Jersey about six miles northeast of Trenton, and stretching southwestward, occupies the southern border of Pennsylvania, extending into Delaware and Maryland. It is subdivided by Rogers into three belts; the first or southernmost of these, passing through Philadelphia, consists of alternations of dark hornblendic and highly micaceous gneiss, with abundance of mica-slate, sometimes coarse-grained, and at other times so fine-grained as to constitute a sort of whet-slate. To the northwestward the strata become still more micaceous, with garnets and beds of hornblende slate, till we reach the second subdivision, which consists of a great belt of highly talcose and micaceous schists, with steatite and serpentine, and is in its turn succeeded by a third, narrow belt resembling the less micaceous members of the first or southernmost subdivision. The micaceous schists of this region abound in staurolite, garnet, cyanite and corundum, and are traversed by numerous irregular granitic veins containing beryl and tourmaline. All of these characters lead us to refer the gneiss of this southern district to the third or White Mountain series, with the exception of the middle subdivision, which presents the aspect of the second or Green Mountain series.

Above the hypozoic gneisses Rogers has placed his azoic or semi-metamorphic series, which is traceable from the vicinity of Trenton to the Schuylkill, along the northern boundary of the southern hypozoic gneiss district. This series is supposed by Rogers to be an altered form of the primal sandstones and slates, and is described as consisting of a feldspathic quartzite or eurite, containing in some cases porphyritic beds with crystals of feldspar and hornblende, together with various crystalline schists; including in fact the whole of the great serpentine belt of Montgomery, Chester and Lancaster counties, with its steatites, hornblendic, dioritic, chloritic, and micaceous schists (often garnet-bearing), together with a band of argillite, affording roofing-slates. With this great series are associated chromic and titanic iron, and ores of nickel and copper. Veins of albite with corundum also intersect this series near Unionville. We are repeatedly assured by

Rogers that these rocks so much resemble the underlying hypozoic gneiss, as to be readily confounded with them; and when compared with the latter, as displayed in the southern district, it is difficult to believe that we have in this so-called azoic or metamorphic series of the Montgomery and Chester valleys, anything else than a repetition of these same crystalline schists which have been described along their southern boundary, representing the Green Mountain and the White Mountain series. We thus avoid the difficulty of supposing that we have in this region two sets of serpentinic rocks, and two of mica-schists, lithologically similar, but of widely different ages, — a conclusion highly improbable. It should be said that Rogers, in accordance with the notions then generally received, looked upon serpentine as an eruptive rock, which had altered the adjacent strata, converting the mica-schists into steatitic and chloritic rocks.

This so-called azoic series, according to Rogers, underlies the auroral limestone of Pennsylvania, thus apparently occupying the horizon of the primal paleozoic division or Potsdam series. We find, however, in his report on the geology of the state, no satisfactory evidence of the identity of the two series. On the contrary, a very different conclusion would seem to follow from certain facts there detailed. The azoic or so-called metamorphic primal strata are said to have a very uniform nearly vertical dip, or with high angles to the southward, while the micaceous and gneissic strata of the northern subdivision of the southern district of so-called hypozoic rocks, limiting these last to the south, present either minute local contortions or wide gentle undulations, with comparatively moderate dips, for the most part to the northward. * From this, I think we may infer that the nearly vertical strata must be, in truth, older underlying rocks belonging, not to the paleozoic system, but to our second series of crystalline schists. We conclude, then, that while the gneisses to the northwest, and probably those along the southeast rim of the Mesozoic basin of Pennsylvania are Laurentian, the great valley southward to the Delaware is occupied by the rocks of the Green Mountain and White Mountain series. The same two types of rocks, extending to the northeast, are developed about New York city, in the mica-schists of Manhattan and the serpentines of Staten Island and Hoboken;

---

* Rogers, Geology of Pennsylvania, I, pp. 69–74, and 154–158.

while in the range of the Highlands, the gneiss belt of the South Mountain crosses the Hudson river.

The three series of gneissic rocks which we have distinguished in our section to the northward have, in southeastern New York, as in Pennsylvania, been grouped together in the primary system, and may thence all be traced into western New England. In Dr. Percival's geological report and map of Connecticut, published in 1840, it will be seen that he refers to the gneiss of the Highlands two gneissic areas in Litchfield county ; the one occupying parts of Cornwall and Ellsworth, and the other extending from Torrington, northward through Winchester, Norfolk and Colebrooke into Berkshire county, Massachusetts. Farther investigations may confirm the accuracy of Percival's identification, and show the Laurentian age of these New England gneisses, a view which is apparently supported by the mineralogical characters of some of the rocks in this region. Emmons informs us that primary limestones with graphite, (perhaps Laurentian), are met with in the Hoosic range in Massachusetts east of the Stockbridge (Taconic) limestones.

The rocks of the second series are traceable from southwestern Connecticut northward to the Green Mountains in Vermont, and the micaceous schists and gneisses of the third or White Mountain series are found both to the east and the west of the Mesozoic valley in Connecticut and Massachusetts. They also occupy a considerable area in eastern Vermont, where they are separated from the White Mountain range by an outcrop of rocks of the second series. To the southeast of the White Mountains, along our line of section, the same mica-schists and gneisses, often with very moderate dips, extend as far as Portland, Maine, where they are interrupted by the outcropping of greenish chloritic and chromiferous schists, in nearly vertical beds, which appear to belong to the second series.

I find that the strata of the second series appear from beneath the Carboniferous at Newport, Rhode Island, in a nearly vertical attitude, and also in the vicinity of Boston and Brighton, Saugus and Lynnfield. Their relations in this region to the gneisses with crystalline limestones of Chelmsford, etc., which I have referred to the Laurentian series,* have yet to be determined.

We have already mentioned that the crystalline rocks of Pennsylvania pass into Maryland and Virginia, where, as H. D. Rogers

---

* Amer. Jour. Sci., II, xlix, 75.

informs us, they appear in the mountains of the Blue Ridge. It remains to be seen whether the three types which we have pointed out in Pennsylvania are to be recognized in this region. A great belt of crystalline schists extends from Virginia through North and South Carolina, and into eastern Tennessee, where, according to Safford, these rocks underlie the Potsdam. It is easy, from the reports of Lieber on the geology of South Carolina, to identify in this state the two types of the Green Mountain and White Mountain series. The former, as described by him, consists of talcose, chloritic and epidotic schists, with diorites, steatites, actinolite-rock and serpentines. It may be noted that he still adheres to the notion of the eruptive origin of the last three rocks, which the observations of Emmons, Logan and myself in the Green Mountains have shown to be untenable. These rocks in South Carolina generally dip at very high angles. The great gneissic area of Anderson and Abbeville districts is described by Lieber as consisting of fine-grained grey gneisses with micaceous and hornblendic schists, and is cut by numerous veins of pegmatite, holding garnet, tourmaline and beryl. These rocks, which have the characters of the White Mountain series, appear, from the incidental observations to be found in Lieber's reports, to belong to a higher group than the chloritic and serpentinic series, and to dip at comparatively moderate angles.

Professor Emmons, whose attention was early turned to the geology of western New England, did not distinguish between the three types which we have defined, but, like Rogers in Pennsylvania, included all the crystalline rocks of that region in the primary system. It is to him, however, that we owe the first correct notions of the geological nature and relations of the Green Mountains. These, he has remarked, are often made to include two ranges of hills belonging to different geological series. The eastern range, including the Hoosic Mountain in Massachusetts, and Mount Mansfield in Vermont, he referred to the primary; which he described as including gneiss, mica-schist, talcose slate and hornblende, with beds and veins of granite, limestone, serpentine and trap. He declared, moreover, that there is no clear line of demarcation among the various schistose primary rocks, and cited, as an illustration, the passage into each other of serpentine, steatite and talcose schist. His description of the crystalline rocks of this range will be recognized as comprehensive and truthful.

To the west of the hills of primary schist, he placed his Taconic system, named from the Taconic hills, which run from north to south along the boundary line of New York and Massachusetts and form a range parallel with the Green Mountains. The lower portions of the Taconic system, according to Emmons, are schistose rocks made up from the ruins of the primary schists which lie to the east of them. Thus the talcose schists of Berkshire are said to be regenerated rocks, belonging to the newer system, but showing the color and texture of the older talcose schists from which they were formed. How far this is true of these particular strata may be a question, for there is reason to believe that Emmons included among his Taconic rocks some beds belonging to the older crystalline series of the Green Mountains ; yet it is not less true that the possibility of derived rocks of this kind is one which has been too much overlooked by geologists. Emmons elsewhere remarks that while the talcose slates of the primary are associated with steatite and with hornblende, these are never found in the Taconic rocks, and also, that epidote, actinolite, titanium (rutile), etc., which are characteristic minerals of the primary, are wanting in the Taconic system.

The statements of Emmons on this point, were sufficiently explicit ; he included in the primary system all of the crystalline schists of the Green Mountains, except certain talcose and micaceous beds, which he supposed to be composed from the ruins of similar strata in the primary, and to constitute, with a great mass of other rocks, the Taconic system ; which was, in its turn, unconformably overlaid by the Potsdam sandstone and Calciferous sandrock of the New York system. His views have, however, been misunderstood by more than one of his critics ; thus, Mr. Marcou, while defending the Taconic system, makes it to include the three groups just mentioned, viz. : I, the Green Mountain gneiss ; II, the Taconic strata as defined by Emmons, and III, the Potsdam sandstone,* thus uniting in one system the crystalline schists and the overlying uncrystalline fossiliferous sediments, in direct opposition to the plainly expressed teachings of Emmons, as laid down in his report on the Geology of the Northern District of New York, and later, in 1846,† in his work on the Taconic system.

In the geological survey of the state of New York, the rocks of

---

* Proc. Bost. Nat. Hist. Soc., Nov. 6, 1861, and Amer. Jour. Sci., II, xxxiii, 282.
† Loc. cit., p. 139, and Agricult., N. York, I, 53.

the Champlain division, including the strata from the base of the Potsdam sandstone to the summit of the Loraine or Hudson River shales, had, by his colleagues, been looked upon as the lowest of the paleozoic system. Professor Emmons, however, was led to regard the very dissimilar strata of the Taconic hills as constituting a distinct and more ancient series. A similar view had been held by Eaton, who placed, as we have already seen, above the crystalline schists of the Green Mountains, his primary quartzose and calcareous formations, followed to the westward by transition argillites and sandstones, which latter appear to have corresponded to the Potsdam sandstone of New York. Emmons, however, gave a greater form and consistency to this view, and endeavored to sustain it by the evidence of fossils, as well as by structure. The Taconic system, as defined by him, may be briefly described as a series of uncrystalline fossiliferous sediments reposing unconformably on the crystalline schists of the Green Mountains, and partly made up of their ruins; while it is, at the same time, overlaid unconformably by the Potsdam and Calciferous formations of the Champlain division, and constitutes the true base of the paleozoic column,—thus occupying the position of the British Cambrian.

Although he claimed to have traced this Taconic system throughout the Appalachian chain from Maine to North Carolina, it is along the confines of Massachusetts and New York that its development was most minutely studied. He divided it into a lower and an upper division, and estimated its total thickness at not less than thirty thousand feet, consisting, in the order of deposition, of the following members:—1. Granular quartz; 2. Stockbridge limestone; 3. Magnesian slate; 4. Sparry limestone; 5. Roofing-slate, graptolitic; 6. Silicious conglomerate; 7. Taconic slate; 8. Black slate. The apparent order of superposition differs from this, and it was conceived by Professor Emmons that during the accumulation of these Taconic rocks, the Green Mountain gneiss, which formed the eastern border of the basin, was gradually elevated so as to bring successively the older members above the ocean from which the sediments were being deposited. From this it resulted that the upper members of the system, such as the black slates, were confined to a very narrow belt, and never extended far eastward; although he admits that denudation may have removed large portions of these upper beds. At a subsequent period, a series of parallel faults, with upthrows on the eastern side, is supposed to have broken the strata, given them an eastward dip, and caused the newer beds to pass successively beneath the older ones, thus producing an *apparently inverted succession*, and making their present seeming order of superposition completely deceptive. In speaking of this supposed arrangement of the members of his Taconic system, Emmons alluded to them as "inverted strata;" while by Mr. Marcou, the strata were said to be "overturned on each side of the crystalline and eruptive rocks which occupy the centre of the chain, producing thus a fan-shaped structure," etc.* I have elsewhere shown that this notion, though to some extent countenanced by his vague and inaccurate use of terms, was never entertained by Emmons, whose own view, as defined in his *Taconic System* (p. 17),† is that just explained.

The view of Emmons that there exists at the western base of the Green Mountains, older fossiliferous series underlying the Potsdam, met with general opposition from American geologists. In May, 1844, H. D. Rogers, in his address as President, before the American Association of Geologists, then met at Washington, criticised this view at length, and referred to a section from Stockbridge, Massachusetts, to the Hudson River, made by W. B. Rogers and himself, and by them laid before the American Philosophical Society in January, 1841. They then maintained that the quartz-rock of the Hoosic range was Potsdam, the Berkshire marble identical with the blue limestone of the Hudson valley, and the associated micaceous and talcose schists, altered strata of the age of the slates at the base of the Appalachian system; that is to say, primal in the nomenclature of the Pennsylvania survey.

In 1843 Mather had asserted the Champlain age of the same crystalline rocks, and claimed that the whole of the division was there represented, including the Potsdam, the Hudson River group, and the intermediate limestones. ‡ The conclusion of Mather was cited with approbation by Rogers, who apparently adopted it, and

---

* Comptes Rendus de l'Acad., LIII, 804.

† See my farther discussion of the matter, Amer. Jour. Sci., II, xxxii, 427, xxxiii, 135, 281. It is by an oversight that I have, in the latter volume, page 136, represented Barrande as sharing the misconception of Marcou, although his language, without careful scrutiny, would lead us to such a conclusion. In fact in the Bull. Soc. Geol. de France (II, xviii, 231), in an elaborate study of the Taconic question, Barrande heads a section thus, "*Renversement conçu pour tout un système*," and then proceeds to show that the *renversement* or *overturn* is only apparent, by explaining, in the language of Emmons, the view already set forth above.

‡ Geology of the Southern District of New York, p. 438.

claimed that Hitchcock held a similar view. It will be seen that these geologists thus united in one group, the schists of the Hoosic range (regarded by Emmons as primary), with those of the Taconic range, and referred both to the age of the Champlain division, the whole of which was supposed to be included in the group.

In the same address Professor Rogers raised a very important question. Having referred to the Potsdam sandstone, which on Lake Champlain forms the base of the paleozoic system, he inquires, " Is this formation then the lowest limit of our Appalachian masses generally, or is the system expanded downward in other districts by the introduction beneath it of other conformable sedimentary rocks?" He then proceeded to state that from the Susquehanna River, southwestward, a more complex series appears at the base of the lower limestone than to the north of the Schuylkill, and in some parts of the Blue Ridge he includes in the primal division (beneath the Calciferous sandrock) " at least four independent and often very thick deposits, constituting one general group. in which the Potsdam or white sandstone (with Scolithus) is the second in descending order." This sandstone is overlaid by many hundred feet of arenaceous and ferriferous fucoidal slate, and underlaid by coarse sandy shales and flagstones ; below which, in Virginia and East Tennessee, is a series of heterogeneous conglomerates, which rest on a great mass of crystalline strata. The accuracy of these statements is confirmed by Safford, who, in his recent report on the geology of Tennessee (1869), places at the base of the column a great series of crystalline schists, apparently representatives of those of southeastern Pennsylvania. Upon these repose what Safford designates as the Potsdam group, including, in ascending order, the Ococee slates and conglomerates, estimated at 10,000 feet, and the Chilhowee shales and sandstones. 2,000 feet or more, with fucoids, worm-burrows and Scolithus. These are conformably overlaid by the Knoxville division, consisting of fucoidal sandstones, shales, and limestones, the latter two holding fossils of the age of the Calciferous sandrock. It is noteworthy that these rocks are greatly disturbed by faults, and that in Chilhowee Mountain the lower conglomerates are brought on the east against the Carboniferous limestone, by a vertical displacement of at least 12,000 feet. The general dip of all these strata, including the basal crystalline schists, is to the southeast.

The primal paleozoic rocks of the Blue Ridge were then by Rog-

ers, as now by Safford, looked upon as wholly of Potsdam age, in
cluding the Scolithus sandstone as a subordinate member, so that
the strata beneath this were still regarded as belonging to the New
York system. Hence, while Rogers inquires whether the Taconic
system "may not along the western border of Vermont and Massachusetts include also some of the sandy and slaty strata here
spoken of as lying beneath the Potsdam sandstone"[*] he would still
embrace these lower strata in the Champlain division.

Thus we see that at an early period the rocks of the Taconic
system were, by Rogers and Mather, referred to the Champlain division of the New York system, a conclusion which has been sus-
tained by subsequent observations. Before discussing these, and
their somewhat involved history, we may state two questions which
present themselves in connection with this solution of the problem.
First, whether the Taconic system, as defined by Emmons, includes
the whole or a part of the Champlain division ; and second, whether it embraces any strata older or newer than the members of this
portion of the New York system. With reference to the first
question it is to be remarked that in their attempts to compare the
Taconic rocks with those of the Champlain division as seen farther
to the west, observers were led by lithological similarities to identify the upper members of the latter with certain portions of the
Taconic. In fact, the Trenton limestone, with the Utica slates
and the Loraine or Hudson River shales, making together the upper
half of the Champlain division (in which Emmons moreover included the overlying Oneida and Medina conglomerates and sandstones), have in New York an aggregate thickness of not less than
three or four thousand feet, and offer many lithological resem-
blances to the great mass of sediments at the western base of the
Green Mountains, to which the name of Taconic had been applied.
It is curious to find that Emmons, in 1842, referred to the Medina
the Red sandrock of the east shore of Lake Champlain, since shown
to be Potsdam ; and, moreover, placed the Sillery sandstone of the
neighborhood of Quebec at the summit of the Champlain division,
as the representative of the Oneida conglomerate ; while at the
same time he noticed the great resemblance which this sandstone,
with its adjacent limestones, bore to similar rocks on the confines
of Massachusetts, already referred by him to the Taconic system.[†]

---

* Amer. Jour. Sci., I, xlvii, 152, 153.
† Geol. Northern District of New York, pp. 124, 125.

This view of Emmons as to the Quebec rocks was adopted by Sir William Logan, when, a few years afterwards, he began to study the geology of that region. The sandstone of Sillery was described by him as corresponding to the Oneida or Shawangunk conglomerate, while the limestones and shales of the vicinity, which were supposed to underlie it, were regarded as the representatives of the Trenton, Utica, and Hudson River formations. * By following these rocks along the western base of the Appalachians into Vermont and Massachusetts, they were found to be a continuation of the Taconic system, which Sir William was thus led to refer to the upper half of the Champlain division, as had already been done by Professor Adams in 1847.† As regards the crystalline strata of the Appalachians in this region, he, however, rejected the view of Emmons, and maintained that put forward by the Messrs. Rogers in 1841, viz., that these, instead of being older rocks, were but these same upper formations of the Champlain division in an altered condition ; a view which was maintained during several years in all of the publications of those connected with the geological survey of Canada.

This conclusion, so far as regards the age of the unaltered fossiliferous rocks from Quebec to Massachusetts, was supposed to be confirmed by the evidence of organic remains found in them in Vermont. Mr. Emmons had described as characteristic of the upper part of the Taconic system, two crustaceans, to which he gave the names of *Atops trilineatus* and *Elliptocephalus asaphoides ;* the other fossils noticed by him being graptolites, fucoids, and what were apparently the marks of annelids. In 1847 Professor James Hall, in the first volume of his Paleontology, declared the Atops of Emmons to be identical with *Triarthrus (Calymene) Beckii*, a characteristic fossil of the Utica slate ; while the Elliptocephalus was referred by him to the genus *Olenus*, now known to belong to the primordial fauna of Sweden, where it is found in slates lying beneath the orthoceratite limestone, and near the base of the paleozoic series. Although, as it now appears, the geological horizon of the Olenus slates was well known to Hisinger, this author in his classic work, *Lethœa Suecica*, published in 1837, represents, by some unexplained error, these slates as overlying the orthoceratite

---

limestone, which is the equivalent of the Trenton limestone of the Champlain division. Hence, as Mr. Barrande has remarked, Hall was justified by the authority of Hisinger's published work in assigning to the Olenus slates of Vermont a position above that limestone, and in placing them, as he then did, on the horizon of the Hudson River or Loraine shales. The double evidence afforded by these two fossil forms in the rocks of Vermont, served to confirm Sir William Logan in placing in the upper part of the Champlain division the rocks which he regarded as their stratigraphical equivalents near Quebec; and which, as we have seen, had some years before been by Emmons himself assigned to the same horizon. The remarkable compound graptolites which occur in the shales of Pointe Levis, opposite Quebec, were described by Professor James Hall in the report of the Geological Survey of Canada for 1857, and were then referred to the Hudson River group; nor was it until August, 1860, that Mr. Billings described from the limestones of this same series at Pointe Levis a number of trilobites, among which were several species of Agnostus, Dikelocephalus, Bathyurus, etc., constituting a fauna whose geological horizon he decided to be in the lower part of the Champlain division.

Just previous to this time, in the Report of the Regents of the University of New York for 1859, Professor Hall had described and figured by the name of Olenus, two species of trilobites from the slates of Georgia, Vermont, which Emmons had wrongly referred to the genus Paradoxides. They were at once recognized by Barrande, who called attention to their primordial character, and thus led to a knowledge of their true stratigraphical horizon, and to the detection of the singular error in Hisinger's book, already noticed, by which American geologists had been misled.* They have since been separated from Olenus, and by Professor Hall referred to a new and closely related genus, which he has named Olenellus, and which is now regarded as belonging to the horizon of the Potsdam sandstone, to which we shall presently advert.

Farther studies of the fossiliferous rocks near Quebec showed the existence of a mass of sediments estimated at about 1200 feet, holding a numerous fauna, and corresponding to a great development of strata about the age of the Calciferous and Chazy formations, or more exactly to a formation occupying a position

---

*For the correspondence on this matter between Barrande, Logan and Hall, see Amer. Jour. Sci., II, xxxi, 210-226.

between these two, and constituting, as it were, beds of passage between them. In this new formation were included the graptolites already described by Hall, and the numerous crustacea and brachiopoda described by Billings, all of which belong to the Levis slates and limestones. To these and their associated rocks Sir William Logan then gave the name of the Quebec group, including, besides the fossiliferous Levis formation, a great mass of overlying slates, sandstones and magnesian limestones, hitherto without fossils, which have been named the Lauzon rocks, and the Sillery sandstones and shales, which he supposed to form the summit of the group, and which had afforded only an Obolella and two species of Lingula ; * the volume of the whole group being about 7000 feet.

The paleontological evidence thus obtained by Billings and by Hall, both from near Quebec and in Vermont, led to the conclusion that the strata of these regions, so much resembling the upper members of the Champlain division, were really a great development, in a modified form, of some of its lower portions. Their apparent stratigraphical relations were explained by Logan by the supposition of "an overturned anticlinal fold, with a crack and a great dislocation running along the summit, by which the Quebec group is brought to overlie the Hudson River group. Sometimes it may overlie the overturned Utica formation, and in Vermont points of the overturned Trenton appear occasionally to emerge from beneath the overlap." He, at the same time, declared that "from the physical structure alone, no person would suspect the break that must exist in the neighborhood of Quebec, and, without the evidence of fossils, every one would be authorized to deny it."†

The rocks from western Vermont, which had furnished to Hall the species of Olenellus, have long been known as the Red sandrock, and as we have seen, were by Emmons, in 1842, referred to the age of the Medina sandstone, a view which the late Professor Adams still maintained as late as 1847. ‡ In the mean time Emmons had, in 1855, declared this rock to represent the Calciferous and Potsdam formations, the brown sandstones of Burlington and Charlotte, Vermont, being referred to the latter.§

---

* See Billings, Paleozoic Fossils of Canada, p. 69.

† Logan's letter to Barrande, Amer. Jour. Sci., II, xxxi, 218. The true date of this letter was December 31st, 1860, but, by a misprint, it is made 1831.

‡ Adams, Amer. Jour. Sci., II, v, 108.

§ Emmons, American Geology, II, 128.

This conclusion was confirmed by Billings, who, in 1861, after visiting the region and examining the organic remains of the Red sandrock, assigned to it a position near the horizon of the Potsdam.* Certain trilobites found in this Red sandrock by Adams in 1847, were by Hall recognized as belonging to the European genus *Conocephalus* (= *Conocephalites* and *Conocoryphe*), whose geological horizon was then undetermined.† The formation in question consists in great part of a red or mottled granular dolomite, associated with beds of fucoidal sandstone, conglomerates and slates. These rocks were carefully examined by Logan in Swanton, Vermont, where, according to him, they have a thickness of 2200 feet, and include toward their base a mass of dark colored shales holding Olenellus with Conocephalites, Obolella, etc.; *Conocephalites Teucer*, Billings, being common to the shales and the red sandy beds.‡ Many of these fossils are also found at Troy and at Bald Mountain, New York, where they accompany the Atops of Emmons, now recognized by Billings as a species of Conocephalites.

A similar condition of things extends northeastward along the Appalachian region. On the south side of the St. Lawrence below Quebec a great thickness of limestones, sandstones, and slates, formerly referred to the Quebec group, is now regarded by Billings as, in part at least, of the Potsdam formation; while on the coast of Labrador, and in northern Newfoundland the same formation, characterized by the same fossils as in Vermont, is largely developed, attaining in some parts, according to Murray, a thickness of 3000 feet or more. Along the northern coast of the island it is nearly horizontal, and appears to be conformably overlaid by about 4000 feet of fossiliferous strata representing the Calciferous sandrock and the succeeding Levis formation.

Mr. Billings has described a section from the Laurentian of Crown Point, New York, to Cornwall, Vermont, from which it appears that to the eastward of a dislocation which brings up the Potsdam to overlie the higher members of the Champlain division, the Potsdam is itself overlaid, at a small angle, by a great mass of limestones representing the Calciferous, and having at the summit some of the characteristic fossils of the Levis formation. Next in

---

*Amer. Jour. Sci., II, xxxii, 232.
†Ibid., II, xxxiii, 374.
‡ Geology of Canada, 1863, p. 281. Amer. Jour. Sci., II, xlvi, 224.

ascending order are not less than 2000 feet of limestones with Trenton fossils (embracing probably the Chazy division), while to the east of this the Levis again appears, including the white Stockbridge limestones. * We have here an evidence that the augmentation in volume observed in the lower members of the Champlain division in the Appalachian region extends to the Trenton, which to the west of Lake Champlain is represented, the Chazy included, by not more than 500 feet of limestone. . The Potsdam, in the latter region, consists of from 500 to 700 feet of sandstone holding Conocephalites and Lingulella, and overlaid by 300 feet of magnesian limestone, the so-called Calciferous sandrock. In the valley of the Mississippi these two formations in Iowa, Missouri, and Texas, are represented by from 800 to 1300 feet of sandstones and magnesian limestones, while in the Black Hills of Nebraska, according to Hayden, the only representative of these lower formations is about one hundred feet of sandstone holding Potsdam fossils.†

In striking contrast to this it has been shown that along the Appalachian range from Newfoundland to Tennessee these lower formations are represented by from 8000 to 15000 feet of fossiliferous sediments. It has been suggested by Logan that these widely differing conditions represent deep-sea accumulations on the one hand, and the deposits from a shallow sea which covered a submerged continental plateau, on the other; the sediments in the two areas being characterized by a similar fauna, though differing greatly in lithological characters and in thickness. To this we may add that the continental area, being probably submerged and elevated at intervals, became overlaid with beds which represent only in a partial and imperfect manner the great succession of strata which were being accumulated in the adjacent ocean. ‡

In a paper which I hope to present to the geological section during the present meeting of the Association it will be shown from a study of the rocks of the Ottawa basin that the typical Champlain division not only presents important paleontological breaks, but evidences of statigraphical discordance at more than one horizon over the continental area, which, as the result of widely spread movements, might be supposed to be represented in the Appalachian region. In the latter Logan has already observed

---

* Amer. Jour. Sci., 227.
† Ibid., II, xxv, 439, xxxi, 234.
‡ Ibid., II, xlvi, 225.

that the absence of all but the highest beds of the Levis along the
eastern limit of the Potsdam, near Swanton, Vermont, while the
whole thickness of them appears a little farther westward, makes
it probable that there is a want of conformity between the two;
and I have in this connection insisted upon the entire absence in
this locality of the Calciferous, which is met with a little farther
south in the section just mentioned, as another evidence of the
same unconformity.* There are also, I think, reasons for sus-
pecting another stratigraphical break at the summit of the Quebec
group, in which case many problems in the geological structure of
this region will be much simplified.

It should be remembered that the conditions of deposition in
some areas have been such that accumulations of strata, corresponding to long geologic periods, and elsewhere marked by stratigraphical breaks, are arranged in conformable superposition; and
moreover that movements of elevation and depression have even
caused great paleontological breaks, which over considerable areas
are not marked by any apparent discordance. Thus the remarkable break in the fauna between the Calciferous and the Chazy is not
accompanied by any noticeable discordance in the Ottawa basin,
and in Nebraska, according to Hayden, the Potsdam, Carboniferous, Jurassic and Cretaceous formations are all represented in
about 1200 feet of conformable strata.† In Sweden the whole
series from the base of the Cambrian to the summit of the Upper
Silurian appears as a conformable sequence, while in North Wales,
although there is no apparent discordance from the base of the
Cambrian to the summit of the Lingula flags, stratigraphical
breaks, according to Ramsay, probably occur both at the base and
the summit of the Tremadoc slates,‡ which are considered equivalent to the Levis formation.

We have seen that, according to Logan, a dislocation a little to
the north of Lake Champlain causes the Quebec group to overlie
the higher members of the Champlain division. The same uplift,
according to him, brings up, farther south, the Red sandrock of
Vermont, which to the west of the dislocation rests upon the up-
turned and inverted strata of various formations from the Calciferous sandrock to the Utica and Hudson River shales. These

---

* Amer. Jour. Sci., II, xlvi, 225.
† Ibid., II. xxv, 440.
‡ Quar. Geol. Journal, xix, page xxxvi.

latter, according to him, are seen to pass for considerable distances beneath nearly horizontal layers of the Red sandrock, the Utica slate, in one case, holding its characteristic fossil, *Triarthrus Beckii.* This relation, which is well shown in a section at St. Albans, figured by Hitchcock,[*] was looked upon by Emmons and by Adams as evidence that the Red sandrock was the representative of the Medina sandstone of the New York system. When, however, the former had recognized the Potsdam age of the sandrock, with its Olenellus, which he supposed to be Paradoxides, this condition of things was conceived to be an evidence of the existence beneath the Potsdam of an older and unconformable fossiliferous series already mentioned.

The objections made by Emmons to Rogers's view of the Champlain age of the Taconic rocks were three-fold : first, the great differences in lithological characters, succession and thickness, between these and the rocks of the Champlain division as previously known in New York; second, the supposed unconformable infraposition of a fossiliferous series to the Potsdam ; and third, the distinct fauna which the Taconic rocks were supposed to contain. The first of these is met by the fact now established that in the Appalachian region, the Champlain division is represented by rocks having, with the same organic remains, very different lithological characters, and a thickness ten-fold greater than in the typical Champlain region of northern New York. The second objection has already been answered by showing that the rocks which pass beneath the Potsdam are really newer strata belonging to the upper part of the division, and contain a characteristic fossil of the Utica slate. As to the third point, it has also been met, so far as regards the Atops and Elliptocephalus, by showing these two genera to belong to the Potsdam formation. If we inquire farther into the Taconic fauna we find that the Stockbridge limestone (the Eolian limestone of Hitchcock), which was placed by Emmons near the base of the Lower Taconic, (while the Olenellus slates are near the summit of the Upper Taconic), is also fossiliferous, and contains, according to the determinations of Professor Hall, species belonging to the genera Euomphalus, Zaphrentis, Stromatopora, Chaetetes and Stictopora.[†] Such a fauna would lead to the con-

---

[*] Geology of Vermont, p. 374.
[†] Geology of Vermont, 419, and Amer. Jour. Sci., II, xxxiii, 419.

clusion that these limestones instead of being older, were really
newer than the Olenellus beds, and that the apparent order of succession was, contrary to the supposition of Emmons, the true one.
This conclusion was still farther confirmed by the evidence obtained in 1868 by Mr. Billings, who found in that region a great
number of characteristic species of the Levis formation, many of
. them in beds immediately above or below the white marbles, *
which latter, from the recent observations of the Rev. Augustus
Wing in the vicinity of Rutland, Vermont, would seem to be
among the upper beds of the Potsdam formation. Thus while
some of the Taconic fossils belong to the Potsdam and Utica
formations, the greater number of them, derived from beds supposed to be low down in the system, are shown to be of the age
of the Levis formation. There is, therefore, at present, no evidence of the existence, among the unaltered sedimentary rocks of
the western base of the Appalachians in Canada or New England,
of any strata more ancient than those of the Champlain division,
to which, from their organic remains, the fossiliferous Taconic
rocks are shown to belong.

Mr. Billings has, it is true, distinguished provisionally what he
has designated an upper and a lower division of the Potsdam, and
has referred to the latter the Red sandrock with the Olenellus
slates of Vermont, together with beds holding similar fossils at
Troy, New York, and along the straits of Bellisle in Labrador and
Newfoundland; the upper division of the Potsdam being represented by the basal sandstones of the Ottawa basin and of the
Mississippi valley.† In the present state of our knowledge of
the local variations in sediments and in their fauna dependent on
depth, temperature and ocean currents, Billings, however, conceives that it would be premature to assert that these two types of
the Potsdam do not represent synchronous deposits.

The base of the Champlain division, as known in the Potsdam
formation of New York, of the Mississippi valley and the Appalachian belt, does not, however, represent the base of the paleozoic
series in Europe. The Alum slates in Sweden are divided into
two parts, an upper or Olenus zone, and a lower or Conocoryphe
zone, as distinguished by Angelin. The latter is characterized by

---

* Amer. Jour. Sci., II, xlvi, 227.
† Report Geol. of Canada, 1863–66, p. 236.

the genus Paradoxides, which also occupies a lower division in the primordial paleozoic rocks of Bohemia (Barrande's stage C), the greater part of which are regarded as the equivalent of the Olenus zone of Sweden and the Potsdam of North America. The Lingula flags of Wales belong to the same horizon, and it is at their base, in strata once referred to the Lower Lingula flags, that the Paradoxides is met with. These strata, for which Hicks and Salter, in 1865, proposed the name of the Menevian group, are regarded as corresponding to the lower division of the Alum slates, and, like it, contain a fauna not yet recognized in the basal rocks of the New York system. We here approach the debatable land between the Cambrian and the Silurian of the British geologists. The Cambrian, as originally claimed by Sedgwick, included in its upper division the Middle and Upper Lingula flags, with the over-lying Tremadoc slates, to the base of the Llandeilo rocks, and may be regarded as equivalent to the Potsdam, Calciferous and Levis formations ; while in the Lower Cambrian were embraced the Lower Lingula flags and the Upper and Lower Longmynd rocks, corres-ponding respectively to the Harlech grits and the Llanberis slates. A portion of the Cambrian has, however, been claimed for the Silurian by Murchison, who draws the dividing line at the top of the Longmynd rocks, leaving the three divisions of the Lingula flags in the Silurian. Lyell, on the contrary, remarks that the Menevian beds, which were, on lithological grounds, made by Sedgwick a part of the Lower Lingula flags, have been shown by Hicks and Salter to be very distinct from these paleontologi-cally ; and, while he includes the Menevian in the Lower Cam-brian, refers the whole of the Lingula flags to the Upper Cambrian.

Lyell therefore admits the whole of the Cambrian system as originally defined by Sedgwick, and the same classification is now adopted by Linarsson, in Sweden, where in Westrogothia, the Cam-brian rocks, (resting unconformably on the crystalline schists to be noticed farther on), are overlaid conformably by the orthoceratite limestones, which are by him regarded as forming the base of the Silurian, and as the equivalent of the Llandeilo rocks of Wales. The total thickness of these lower rocks in Sweden, including the representatives of the Lingula flags, the Menevian beds and an underlying fucoidal (Eophyton) sandstone, is only three hundred feet, while the first two divisions in Wales have a thickness of five to six thousand, and the Harlech grits and Llanberis slates (including the Welsh roofing-slates beneath) amount to eight thousand feet additional. Recent researches show that these lower rocks in Wales contain an abundant fauna, extending downward some 2800 feet from the Menevian to the very base of strata regarded as the representatives of the Harlech grits. The brachiopoda of the Harlech beds appear identical with those of the Menevian, but new species of *Conocephalites*, *Microdiscus* and *Paradoxides* are met with, besides a new genus, *Plutonia*, allied to the last mentioned. Mr. Hicks, to whom we owe these discoveries,[*] remarks, that the Menevian gives us, for the present, a well marked paleontological horizon for the summit of the Cambrian, corresponding with the Lower Cambrian as defined by Sedgwick.

The Upper Cambrian in North America would thus include the lower half of the Champlain division from the base of the Potsdam to the summit of the Levis (including perhaps the Chazy), while the Lower Cambrian, (the Cambrian of Murchison and Hicks) is represented by the strata holding Paradoxides in Newfoundland, New Brunswick and eastern Massachusetts. Although no strata marked by these fossils have yet been found in the Appalachians, it is not improbable that such may yet be met with. In May, 1861, I called attention to the fact that beds of quartzose conglomerate at the base of the Potsdam in Hemmingford, near the outlet of Lake Champlain on its western side, contain fragments of green and black slates, "showing the existence of argillaceous slates before the deposition of the Potsdam sandstone."[†] The more ancient strata, which furnished these slaty fragments to the Potsdam conglomerate, have perhaps been destroyed, or are concealed, but they or their equivalents may yet be discovered in some part of the great Appalachian region. They should not, however, be called Taconic, but receive the prior designation of Cambrian, unless, indeed, it shall appear that the source of these slate fragments was the more argillaceous beds of the still older Huronian schists. Emmons regarded his Taconic system as the equivalent of the Lower Cambrian of Sedgwick, but when in 1842, Murchison announced that the name of Cambrian had ceased to have any zoological significance, being identical with Lower Silurian,[‡] Emmons, conceiving, as he tells us, that all

---

[*] Geol. Mag., V, 306; and Rep. Brit. Assoc., 1868, p. 69; also Harkness and Hicks in Nature, Proc. Geol. Soc., May 10, 1871.

[†] Amer. Jour. Sci., II, xxxi, 404.

[‡] Proc. Geol. Soc., London, III, 642.

Cambrian rocks were not Silurian, instead of maintaining Sedgwick's name, which with the progress of paleontological study is assuming a great zoological importance, devised the name of Taconic, as synonymous. with Lower Cambrian ;* although, as we have seen, there is as yet no paleontological evidence to identify any portion of the Taconic strata with the well-defined Lower Cambrian rocks of our eastern shores.

The crystalline infra-Silurian strata, to which the name of the Huronian series has been given by the Geological Survey of Canada, have sometimes been called Cambrian from their resemblance to certain rocks in Anglesea, which have been looked upon as altered Cambrian. The typical Cambrian rocks of Wales, down to their base, are however uncrystalline sediments, and, as pointed out by Dr. Bigsby in 1863,† are not to be confounded with the Huronian, which he regarded as equivalent to the second division of the so-called azoic rocks of Norway, the *Urschiefer* or primitive schists, which in that country rest unconformably on the primitive gneiss (*Urgneiss*), and are in their turn overlaid unconformably by the fossiliferous Cambrian strata. This second or intermediate series in Norway is characterized by eurites, micaceous, chloritic and hornblendic schists, with diorites, steatite and dark colored serpentines, generally associated with chrome ; and abounds in ores of copper, nickel and iron. In its mineralogical and lithological characters, the Urschiefer corresponds with what we have designated the second series of crystalline schists. It is, in Norway, divided into a lower or quartzose division, marked by a predominance of quartzites, conglomerates and more massive rocks, and an upper and more schistose division. Macfarlane, who was familiar with the rocks of Norway, after examining both the Huronian of Lake Superior and the crystalline strata of the Green Mountains, had already, in 1862, declared his opinion that both of these were representatives of the Norwegian Urschiefer, ‡ thus anticipating, from his comparative studies, the conclusions of Bigsby.

The crystalline rocks of Anglesea and the adjacent part of Caernarvon, which have been described and mapped by the British Geological Survey as altered lowest Cambrian, are directly overlaid by strata of the Llandeilo and Bala divisions, corresponding

---

* Emmons, Geol. N. District of New York, 162; and Agric. of New York, I, 49.
† Quar. Jour. Geol. Soc., XIX, 36.
‡ Canadian Naturalist, VII, 125.

to the Trenton and Hudson River formations. If we consult Ramsay's report on the region, it will be found that he speaks of them as "probably Cambrian," and states as a reason for that opinion, that they are connected by certain beds of intermediate lithological characters with strata of undoubted Cambrian age.* These, however, as he admits, present great local variations, and, after carefully scanning the whole of the evidence adduced, I am inclined to see in it nothing more than the existence, in this region, of Cambrian strata made up from the ruins from the great mass of pre-Cambrian schists, which are the crystalline rocks of Anglesea. Such a phenomenon is repeated in numerous instances in our North American rocks, and is the true explanation of many supposed examples of passage from crystalline schists to uncrystalline sediments. The Anglesea rocks are a highly inclined and much contorted series of quartzose, micaceous, chloritic and epidotic schists, with diorites and dark colored chromiferous serpentines, all of which, after a careful examination of them in the collections of the Geological Survey of Great Britain, appear to me identical with the rocks of the Green Mountain or Huronian series. A similar view of their age is shared by Phillips and by Sedgwick, in opposition to the opinion of the British survey. The former asserts that the crystalline schists of Anglesea are "below all the Cambrian rocks ;"† while Sedgwick expresses the opinion that they are of "a distinct epoch from the other rocks of the district, and evidently older."‡

Associated with the fossiliferous Devonian rocks of the Rhine, is a series of crystalline schists, similar to those just noticed, seen in the Taunus, the Hundsrück and the Ardennes. These, in opposition to Dumont, who regarded them as belonging to an older system, are declared by Römer to have resulted from a subsequent alteration of a portion of the Devonian sediments. §

Turning now to the Highlands of Scotland, we have a similar series of crystalline schists, presenting all the mineralogical characters of those of Norway and of Anglesea, which, according to Murchison and Giekie, are neither of Cambrian nor pre-Cambrian age, but are younger than the fossiliferous limestones of the west-

---

* Geol. of North Wales, pp. 145, 175.
† Manual of Geology (1855) 89.
‡ Geol. Journal for 1845, 449.
§ Naumann, Geognosie, 2d edition, II, 383.

ern coast (about the horizon of the Levis formation) which seem to pass beneath them. Professor Nicol, on the contrary, maintains that this apparent super-position is due to uplifts, and that these crystalline schists are really older than either Cambrian or Silurian, both of which appear to the west of them as uncrystalline sediments, resting on the Laurentian. He does not, however, confound these crystalline schists of the Scottish Highlands with the Laurentian, from which they differ mineralogically, but regards them as a distinct series.* In the presence of the differences of opinion which have been shown in this controversy, we may be permitted to ask whether, in such a case, stratigraphical evidence alone is to be relied upon. Repeated examples have shown that the most skilful stratigraphists may be misled in studying the structure of a disturbed region where there are no organic remains to guide them, or where unexpected faults and overslides may deceive even the most sagacious. I am convinced that in the study of the crystalline schists, the persistence of certain mineral characters must be relied upon as a guide, and that the language used by Delesse, in 1847, will be found susceptible of a wide application to crystalline strata. "Rocks of the same age have most generally the same chemical and mineralogical composition, and reciprocally, rocks having the same chemical composition and the same minerals, associated in the same manner, are of the same age."†

In this connection the testimony of Professor James Hall is to the point. Speaking of the crystalline schists of the White Mountain series, he says : —

"Every observing student of one or two years experience in the collection of minerals in the New England States, knows well that he may trace a mica-schist of peculiar but varying character from Connecticut, through central Massachusetts, and thence into Vermont and New Hampshire, by the presence of staurolite and some other associated minerals, which mark with the same unerring certainty the geological relations of the rock as the presence of *Pentamerus oblongus*, *P. galeatus*, *Spirifer Niagarensis*, or *S. macropleura*, and their respectively associated fossils do the relations of the several rocks in which these occur." ‡

---

* Quar. Jour. Geol. Soc.; Murchison, XV, 353; Giekie, XVII, 171; Nicol, XVII, 58, XVIII, 443.
† Bull. Soc. Geol. de Fr. (2), IV, 786.
‡ Paleontology of New York, Vol. III, Introduction, page 93.

I am convinced that these crystalline schists of Germany, Anglesea, and the Scotch Highlands, will be found, like those of Norway, to belong to a period anterior to the deposition of the
Cambrian sediments, and will correspond with the newer gneissic
series of our Appalachian region. There exists, in the Highlands
of Scotland, a great volume of fine-grained, thin-bedded mica-
schists with andalusite, staurolite and cyanite, which are met with
in Argyleshire, Aberdeenshire, Banffshire and the Shetland Isles.
Rocks regarded by Harkness as identical with these of the Scottish
Highlands also occur in Donegal and Mayo in Ireland. Through
the kindness of the Rev. Prof. Haughton of Trinity College, and
Mr. Robert H. Scott, then of Dublin, I received some years since,
a large collection of the crystalline rocks of Donegal, which
I am thus enabled to compare with those of North America, and
to assert the existence in the northwest of Ireland, of our second
and third series of crystalline schists. The Green Mountain rocks
are there exactly represented by the dark colored chromiferous
serpentines of Aghadoey, and the steatite, crystalline talc and
actinolite of Crohy Head; while the mica-schist of Loch Derg,
with white quartz, blue cyanite, staurolite and garnet, all united
in the same fragment, cannot be distinguished from specimens
found at Cavendish, Vermont, and Windham, Maine. The fine-
grained andalusite-schists of Clooney Lough are exactly like
those from Mount Washington; while the granitoid mica-slates
from several other localities in Donegal are not less clearly of the
type of the White Mountain series. Similar micaceous schists,
with andalusite (chiastolite), occur on Skiddaw, in Cumberland,
England, the relations of which have been clearly defined by Sedgwick, who groups the rocks of Skiddaw into four divisions. The
lowest of these, succeeding the granite, is a series of crystalline
rocks, not described lithologically, with mineral veins, " having
some resemblance to the rocks of Cornwall," and including
towards the summit, " chiastolite schists and chiastolite rocks."
These are followed in ascending order by two great series of slates
and grits, succeeded by a fourth division of schists, sometimes
carbonaceous, holding in parts fucoids and graptolites, which are
apparently overlaid discordantly by sundry trappean conglomerates and chloritic slates.* The graptolites of the Skiddaw slates

---

* Synopsis of British Paleozoic Rocks, p. lxxxiv, being an introduction to McCoy's
Brit. Pal. Fossils (1855).

are found to be identical with those of the Levis formation,* and it is worthy of notice that although Sedgwick places the mica-schists with andalusite (chiastolite) so far below the graptolitic beds, he elsewhere, in comparing the rocks of North Wales and Cumberland, states that the chloritic and micaceous rocks of Anglesea and Caernarvon are not represented in Cumberland. being distinct from the other rocks of North Wales, and much older.†

In Victoria, Australia, the position of the chiastolite schists, according to Selwyn, is beneath the graptolitic slates. Boblaye, it is true, asserted in 1838 that the chiastolite schists of Les Salles. near Pontivy in Brittany, include *Orthis* and *Calymene*,‡ but when we remember that even experienced observers in the White Mountains for a time mistook for remains of crustacea and brachiopods. certain obscure forms, which they afterwards found not to be organic, and that Dana, in this connection, has called attention to the deceptive resemblance to fossils presented by some imperfectly developed chiastolite crystals in the same region, § we may well require a verification of Boblaye's observation, especially since we find that more recently D'Archiac and Dalimier agree with De Beaumont and Dufrenoy in placing the chiastolite schists of Brittany at the very base of the transition sediments, marking the summit of the crystalline schists. ‖

With regard to the crystalline schists of Lakes Huron and Superior, to which the name of the Huronian system has been given. the observations of all who have studied the region concur in placing them unconformably beneath the sediments which are supposed to represent the base of the New York system, while, on the other hand, they rest unconformably on the Laurentian gneiss, fragments of which are included in the Huronian conglomerates. The gneissic series of the Green Mountains had, however, as we have seen, been. since 1841, regarded by the brothers Rogers, Mather, Hall, Hitchcock, Adams, Logan, myself and others, as of Silurian age. Eaton and Emmons had alone claimed for it a pre-Cambrian age until, in 1862, Macfarlane ventured to unite it with the Huronian system.

---

* Harkness and Salter, Quar. Jour. Geol. Soc., xix, 135.
† Geol. Journal (1845), IV, 583.
‡ Bull. Soc. Geol. de Fr., X, 227.
§ Amer. Jour. Sci., II, i, 415, v, 116.
‖ Bull. Soc. Geol. de Fr., II, xviii, 664.

and to identify both with the crystalline schists of a similar age in Norway. Later observations in Michigan justify still farther this comparison, for not only the more schistose beds of the Green Mountain series, but even the mica-schists of the third or White Mountain series, with staurolite and garnet, are represented in Michigan, as appears by the recent collections of Major Brooks, of the Geological Survey of Michigan, kindly placed in my hands for examination. He informs me that these latter schists are the highest of the crystalline strata in the northern peninsula.

To the north of Lake Superior, as I have already shown elsewhere, the schists of this third series, which, as early as 1861, I compared to those of the Appalachians, are widely spread; while in Hastings County, forty miles north of Lake Ontario, rocks having the mineralogical and lithological characters both of the second and third series are found resting on the first or Laurentian, the three apparently unconformable, and all in turn overlaid by horizontal Trenton limestone.*

We have shown, that in Pennsylvania, while some of these schists of the second and third series were regarded as altered primal rocks by H. D. Rogers, others, lithologically similar, were referred by him to the older so-called azoic series, which we believe to be their true position. Professor W. B. Rogers has lately informed me that in Virginia the gneissic series having the characters of the Green Mountain rocks, is clearly overlaid unconformably by the lowest primal paleozoic strata of the region. Coming northward, the uncrystalline argillites and sandstones holding Paradoxides at Braintree, Massachusetts,† and St. John, New Brunswick, overlie unconformably crystalline schists of the second series, and in the latter region, in one locality, rocks which are by Bailey and Matthew regarded of Laurentian age. In Newfoundland, in like manner, a great series of crystalline schists, in which Mr. Murray recognizes the Huronian system as first studied and described by him in the west, is unconformably overlaid by a group of sandstones, limestones, and slates holding Paradoxides. The peculiar gneisses and mica-schists of the White Mountain series appear to be developed to a great extent in Newfoundland, which has led me to propose for them the name of the Terranovan system. ‡

---

* Amer. Jour. Sci., II, xxxi, 395, and 1, 85.
† Hunt, Proc. Bost. Nat. Hist. Soc., Oct., 19, 1870.
‡ Amer. Jour. Sci., II, 1, 87.

From the part which the ruins of these rocks play in the production of succeeding sediments it is not always easy to define the limits between the ancient mica-schists and the Cambrian strata in these northeastern regions. It is not impossible that the two may graduate into each other, as some have supposed, in Newfoundland and Nova Scotia, but until farther light is thrown upon the subject I am disposed to regard the relation between the two as one of derivation rather than of passage.

We have already alluded to the history of the rocks of the White Mountains, formerly looked upon as primary, and by Jackson described as an old granitic and gneissic axis uplifting the more recent Green Mountain rocks. Their manifest differences from the more ancient gneiss of the Adirondacks, and their apparent superposition to the Green Mountain series, then regarded by the Messrs. Rogers as belonging to the Champlain division, led them in 1846 to look upon the White Mountains as altered strata belonging to the Levant division of their classification, corresponding to the Oneida, Medina and Clinton of the New York system. In 1848 Sir William Logan came to a somewhat similar conclusion. Accepting, as we have seen, the view of Emmons that the strata about Quebec included a portion of the Levant division, and regarding the Green Mountain gneisses as the equivalents of these, he was induced to place the White Mountain rocks still higher in the geological series than the Messrs. Rogers had done, and expressed his belief that they might be the altered representatives of the New York system from the base of the Lower Helderberg to the top of the Chemung; in other words, that they were not Middle Silurian, but Upper Silurian and Devonian. This view, adopted and enforced by me,* was farther supported by Lesley in 1860, and has been generally accepted up to this time. In 1870, however, I ventured to question it, and in a published letter addressed to Professor Dana, concluded from a great number of facts that there exists a system of crystalline schists distinct from, and newer than, the Laurentian and Huronian, to which I gave the provisional name of Terranovan, constituting the third or White Mountain series, which appears not only throughout the Appalachians, but westward to the north of Lake Ontario, and around and beyond Lake Superior. †   Although I have in common with most

---

* Geol. Survey of Canada, Report 1847-48, p. 58; also Amer. Jour. Sci., II, ix, 19.
† Amer. Jour. Sci., II, I, 83.

other American geologists, maintained that the crystalline rocks of the Green Mountain and White Mountain series are altered paleozoic sediments, I find, on a careful examination of the evidence, no satisfactory proof of such an age and origin, but an array of facts which appear to me incompatible with the hitherto received view, and lead me to conclude that the whole of our crystalline schists of eastern North America are not only pre-Silurian but pre-Cambrian in age.

In what precedes, I have endeavored to discuss briefly and impartially some of the points in the history of the older rocks, and of the views which during the past thirty years have been entertained as to their age and geological relations, both in America and in Europe. I have said some things which will provoke criticism, and at the same time, I trust, lead to farther study of these rocks, a correct knowledge of which lies at the basis of geological science.

I cannot, however, conclude this part of my subject without referring to the views put forth in 1869 by Professor Hermann Credner of Leipzig, in an essay on the Eozoic or pre-Silurian formations of North America.* With Macfarlane, he refers to the Huronian the gneissic series of the Green Mountains, but includes with it, as part of the Huronian system, the so-called Lower Taconic rocks of Vermont, "with remains of annelids and crinoids." Credner thus falls into the very error against which Emmons warned American geologists, namely, the confounding in one system the ancient crystalline schists with the newer fossiliferous sediments. Resting unconformably on these, he places, first, the Upper Taconic, corresponding, according to him, to a part of the Quebec group, and second, the Potsdam sandstone. In this he has copied, for the most part, Marcou, who, however, groups the whole of these various divisions in the Taconic system, while Credner, rejecting the name, unites a portion of the Taconic of Emmons with the Huronian system, and refers the other portion, together with the Potsdam, to the Silurian. These same views are set forth in a more recent paper, by the same author, on the Alleghany system, which is accompanied with sections and a geologically colored map.† In this, not content with including in the Huronian both the fossiliferous strata of the Levis formation and

---

* Die Gliederung der Eozoischen Formationsgruppe, u. s. w., pp. 53. Halle, 1869.
† Petermann's Geographische Mittheilungen. 2 Heft, 1871.

the crystalline schists of the Green Mountains, he refers the
gneisses and mica-schists of the White Mountains to the same
system; while the broad area of similar rocks from their base to
the sea at Portland, is regarded as Laurentian. This, on Credner's
map, is also made to include, with the exception of the White
Mountains themselves, all the rocks of the third or White Mountain series which cover so large a part of New England. Those
who have followed the historical sketch already given, can see how
widely these notions of Credner differ from those of Emmons, and
from all other American geologists, and how much they are at
variance with the present state of our knowledge. It is much to
be regretted that so good a geologist and lithologist should, from
a too superficial study, have fallen into these errors, which can
only retard the progress of comparative geognosy, for which he
has done so much. In England, again, Credner confounds the
Cambrian and Huronian, referring to the latter system the whole
of the Longmynd rocks with their characteristic Cambrian fauna,
a view which is supported only by the conjectured Cambrian age
of the crystalline schists of Anglesea, which are probably pre-
Cambrian and veritably Huronian, like the Urschiefer of Scan-
dinavia; which Credner correctly refers to the latter system, as
Macfarlane and Bigsby had done before him. He, moreover, recognizes in the similar crystalline schists of Scotland, the Urals,
and various parts of Germany, including those of Bavaria and
Bohemia, a newer system, overlying the primary or Laurentian
gneiss, and corresponding to the Huronian or Green Mountain
series of North America, while he suggests a correspondence with
similar rocks in Japan, Bengal, and Brazil. In a collection of
rocks brought from the latter country by Professor C. F. Hartt, I
have found, as elsewhere stated,* what appear to be representatives of the three types of crystalline schists which have been
distinguished in eastern North America.

It will be noticed that I have not, in the preceding pages,
referred to the Labradorian (Upper Laurentian) system, which is
characterized by a great predominance of norites and hyperites.
Although occupying a considerable area in the Adirondack region,
it is not certainly known in the Appalachian range, and was,
therefore, omitted in the discussion. In addition to the facts

---

* The Nation, Dec. 1, 1870, and Hartt's Geology of Brazil, p. 550.

given by me in 1869,* it may be added that the observations of
Mr. Richardson, during that season, on the north side of the Gulf
of St. Lawrence, confirm the previous conclusions, and show that
the rocks of the Labradorian (or rather Norian) system there repose transgressively, and often at comparatively moderate angles,
on the nearly vertical Laurentian gneisses.† We may, I think, in
the present state of our knowledge, regard these norites or Norian
rocks as portions of a pre-Huronian system.

II.   *The Origin of Crystalline Rocks.*

We now approach the second part of our subject, namely, the
genesis of the crystalline schists whose history we have just discussed.   The origin of the mineral silicates which make up a great
portion of the crystalline rocks of the earth's surface is a question of much geological interest, which has been to a great degree
overlooked.   The gneisses, mica-schists and argillites of various
geological periods do not differ very greatly in chemical constitution from modern mechanical sediments, and are now very generally regarded as resulting from a molecular re-arrangement of
similar sediments formed in earlier times by the disintegration
of previously existing rocks not very unlike them in composition ;
the oldest known formations being still composed of crystalline
stratified deposits presumed to be of sedimentary origin.   Before'
these the imagination conceives yet earlier rocks, until we reach
the surface of unstratified material which the globe may be supposed to have presented before water had begun its work.   It is
not, however, my present plan to consider this far-off beginning of
sedimentary rocks, which I have elsewhere discussed. ‡

Apart from the clay and sand-rocks just referred to, whose composition may be said to be essentially quartz and aluminous silicates,
chiefly in the forms of feldspars and micas, or the results of their
partial decomposition and disintegration, there is another class
of crystalline silicated rocks which, though far less important in
bulk than the last, is of great and varied interest to the lithologist, the mineralogist, the geologist and chemist.   The rocks of
this second class may be defined as consisting in great part of the
silicates of the protoxyd bases, lime, magnesia and ferrous oxyd,

---

* On Norites, etc., Amer. Jour. Sci., II, xlviii, 180.
† Geol. Survey of Canada, Report 1866-69, p. 306.
‡ Amer. Jour. Science, II, 1, 25.

either alone, or in combination with silicates of alumina and alkalies. They include the following as their chief constituent mineral species:—pyroxene, hornblende, olivine, serpentine, talc, chlorite, epidote, garnet and triclinic feldspars such as labradorite. The great types of this second class are not less well defined than the first, and consist of pyroxenic and hornblendic rocks, passing into diorites, diabases, ophiolites and talcose, chloritic and epidotic rocks. Intermediate varieties resulting from the association of the minerals of this class with those of the first, and also with the materials of non-silicated rocks, such as limestones and dolomites, show an occasional blending of the conditions under which these various types of rocks were formed.

The distinctions just drawn between the two great divisions of silicated rocks, are not confined to stratified deposits, but are equally well marked in eruptive and unstratified masses, among which the first type is represented by trachytes and granites, and the second, by dolerites and diorites. This fundamental difference between acid and basic rocks, as the two classes are called, finds its expression in the theories of Phillips, Durocher and Bunsen, who have deduced all silicated rocks from two supposed layers of molten matter within the earth's crust, consisting respectively of acid and basic mixtures ; the trachytic and pyroxenic magmas of Bunsen. From these, by a process of partial crystallization and eliquation, or by commingling in various proportions, those eruptive rocks which depart more or less from the normal types, are supposed by the theorists of this school to be generated.* The doctrine that these eruptive rocks are not derived directly from a hitherto uncongealed nucleus, but are softened and crystallized sediments, in fact that the whole of the rocks at present known to us have at one time been aqueous deposits, has, however, found its advocates. In support of this view, I have endeavored to show that the natural result of forces constantly in operation, tends to resolve the various igneous rocks into two classes of sediments, in which the two types are, to a great extent, preserved. The mechanical and chemical agencies which transform the crystalline rocks into sediments, separate these more or less completely into coarse, sandy, permeable beds on the one hand, and fine clayey impervious muds on the other. The action of infiltrating atmospheric waters on the first and more silicious strata, removes from them lime, magnesia, iron-oxyd and

---

* Hunt on Some Points of Chemical Geology, Quar. Jour. Geol. Soc., XV, 489.

soda, leaving behind silica, alumina and potash — the elements of granitic, gneissic and trachytic rocks. The finer and more aluminous sediments, including the ruins of the soft and easily abraded silicates of the pyroxene group, resisting the penetration of the water, will, on the contrary, retain their alkalies, lime, magnesia and iron, and thus will have the composition of the more basic rocks. *

A little consideration will, however, show that this process, although doubtless a true cause of differences in the composition of sedimentary rocks, is not the only one, and is inadequate to explain the production of many of the varieties of stratified silicated rocks. Such are serpentine, steatite, hornblende, diallage, chlorite, pinite and labradorite, all of which mineral species form rock-masses by themselves, frequently almost without admixture. No geological student will now question that all of these rocks occur as members of stratified formations. Moreover, the manner in which serpentines are found interstratified with steatite, chlorite, argillite, diorite, hornblende and feldspar rocks, and these, in their turn, with quartzites and orthoclase rocks, is such as to forbid the notion that these various materials have been deposited, with their present composition, as mechanical sediments from the ruins of preexisting rocks ; a hypothesis as untenable as that ancient one which supposed them to be the direct results of plutonic action.

There are, however, two other hypotheses which have been proposed to explain the origin of these various silicated rocks, and especially of the less abundant, and, as it were, exceptional species just mentioned. The first of these supposes that the minerals of which they are composed, have resulted from an alteration of previously existing minerals, often very unlike in composition to the present, by the taking away of certain elements and the addition of certain others. This is the theory of metamorphism by pseudomorphic changes, as they are called, and is the one taught by the now reigning school of chemical geologists, of which the learned and laborious Bischof, whose recent death science deplores, may be regarded as the great exponent. The second hypothesis supposes that the elements of these various rocks were originally deposited as, for the most part, chemically formed sediments, or precipitates ; and that the subsequent changes have been simply

---

Quar. Jour. Geol. Soc., xv, 489; also, Amer. Jour. Sci., II, xxx, 133.

molecular, or, at most, confined in certain cases to reactions between the mingled elements of the sediments, with the elimination of water and carbonic acid. It is proposed to consider briefly, these two opposite theories, which seek to explain the origin of the rocks in question respectively by pseudomorphic changes in preëxisting crystalline rocks, and by the crystallization of aqueous sediments, for the most part chemically formed precipitates.

Mineral pseudomorphism, that is to say, the assumption by one mineral substance of the crystalline form of another, may arise in several ways. First of these is the filling up of a mould left by the solution or decomposition of an imbedded crystal, a process which sometimes takes place in mineral veins, where the processes of solution and deposition can be freely carried on. Allied to this, is the mineralization of organic remains, where carbonate of lime or silica, for example, fills the pores of wood. When subsequent decay removes the woody tissue, the vacant spaces may, in their turn, be filled by the same or another species. * In the second place, we may consider pseudomorphs from alteration, which are the result of a gradual change in the composition of a mineral species. This process is exemplified in the conversion of feldspar into kaolin by the loss of its alkali and a portion of silica, and the fixation of water, or in the change of chalybite into limonite by the loss of carbonic acid and the absorption of water and oxygen.

The doctrine of pseudomorphism by alteration as taught by Gustaf Rose, Haidinger, Blum, Volger, Rammelsberg, Dana, Bischof, and many others, leads them, however, to admit still greater and more remarkable changes than these, and to maintain the possibility of converting almost any silicate into any other. Thus, by referring to the pages of Bischof's Lehrbuch der Geognosie, it will be found that serpentine is said to exist as a pseudomorph after augite, hornblende, olivine, chondrodite, garnet, mica, and probably also after labradorite, and even orthoclase. Serpentine rock or ophiolite is supposed to have resulted, in different cases, from the alteration of hornblende-rock, diorite, granulite and even granite. Not only silicates of protoxyds and aluminous silicates are conceived to be capable of this transformation, but probably also quartz itself; at least, Blum asserts that meerschaum, a closely re-

---

* Hunt on the Silicification of Fossils, Canadian Naturalist, new series, I, 46.

lated silicate of magnesia, which sometimes accompanies serpentine, results from the alteration of flint ; while according to. Rose, serpentine may even be produced from dolomite, which we are told is itself produced by the alteration of limestone. But this is not all, — feldspar may replace carbonate of lime, and carbonate of lime, feldspar, so that, according to Volger, some gneissoid limestones are probably formed from gneiss by the substitution of calcite for orthoclase. In this way, we are led from gneiss or granite to limestone, from limestone to dolomite, and from dolomite to serpentine, or more directly from granite, granulite or diorite to serpentine at once, without passing through the intermediate stages of limestone and dolomite, till we are ready to exclaim in the words of Goethe : —

> "Mich ängstigt das Verfängliche
> Im widrigen Geschwätz,
> Wo Nichts verharret, Alles flieht,
> Wo schon verschwunden was man sieht,"*

which we may thus translate : — " I am vexed with the sophistry in their contrary jargon, where nothing endures, but all is fugitive, and where what we see has already passed away."

By far the greater number of cases on which this general theory of pseudomorphism by a slow process of alteration in minerals, has been based are, as I shall endeavor to show, examples of the phenomenon of mineral envelopment, so well studied by Delesse in his essay on Pseudomorphs,† and may be considered under two heads : — first, that of symmetrical envelopment, in which one mineral species is so enclosed within the other that the two appear to form a single crystalline individual. Examples of this are seen when prisms of cyanite are surrounded by staurolite, or staurolite crystals completely enveloped in those of cyanite, the vertical axes of the two prisms corresponding. Similar cases are seen in the enclosure of a prism of red in an envelope of green tourmaline, of allanite in epidote, and of various minerals of the pyroxene group in one another. The occurrence of muscovite in lepidolite, and of margarodite in lepidomelane, or the inverse, are well-known examples, and, according to Scheerer, the crystallization of serpentine around a nucleus of olivine is a similar case. This phenomenon of symmetrical envelopment, as remarked by Delesse, shows

---

*Chinesisch-Deutsche Jahres und Tages Zeiten, xi.
†Annales des Mines, V, xvi, 317–392.

itself with species which are generally isomorphous or homœomorphous, and of related chemical composition. Allied to this is the repeated alternation of crystalline laminæ of related species, as in perthite, the crystalline cleavable masses of which consist of thin, alternating layers of orthoclase and albite.

Very unlike to the above are those cases of envelopment in which no relations of crystalline symmetry nor of similar chemical constitution can be traced. Examples of this kind are seen in garnet crystals, the walls of which are shells, sometimes no thicker than paper, enclosing in different cases, crystalline carbonate of lime, epidote, chlorite or quartz. In like manner, crystalline shells of leucite enclose feldspar, hollow prisms of tourmaline are filled with crystals of mica or with hydrous peroxyd of iron, and crystals of beryl with a granular mixture of orthoclase and quartz, holding small crystals of garnet and tourmaline, a composition identical with the enclosing granitic veinstone. * Similar shells of galenite and of zircon, having the external forms of these species, are also found filled with calcite. In many of these cases the process seems to have been first the formation of a hollow mould or skeleton-crystal (a phenomenon sometimes observed in salts crystallizing from solutions), the cavity being subsequently filled with other matters. Such a process is conceivable in free crystals found in veins, as for example, galenite, zircon, tourmaline, beryl and some examples of garnet, but is not so intelligible in the case of those garnets imbedded in mica-schist, studied by Delesse, which enclosed within their crystalline shells irregular masses of white quartz, with some little admixture of garnet. Delesse conceives these and similar cases to be produced by a process analogous to that seen in the crystallization of calcite in the Fontainebleau sandstone ; where the quartz grains, mechanically enclosed in well-defined rhombohedral crystals, equal, according to him, sixty-five per cent., of the mass. Very similar to these are the crystalloids with the form of orthoclase, which sometimes consist in large part of a granular mixture of quartz, mica and orthoclase, with a little cassiterite, and in other cases, contain two-thirds their weight of the latter mineral, with an admixture of orthoclase and quartz. Crystals with the form of scapolite, but made up, in a great part, of mica, seem to be like cases of envelopment, in which a small proportion of one substance in the act of crystallization, compels in-

---

* Report Geol. Survey of Canada, 1866, page 189.

to its own crystalline form a large portion of some foreign material, which may even so mask the crystallizing element that this becomes overlooked, as of secondary importance. The substance which, under the name of houghite, has been described as an altered spinel, is found by analysis to be an admixture of völlknerite with a variable proportion of spinel, which, in some specimens, does not exceed eight per cent., but to which, nevertheless, these crystalloids appear to owe their more or less complete octohedral form. *

The above characteristic examples of symmetrical and asymmetrical envelopment are cited from a great number of others which might have been mentioned. Very many of these are by the pseudomorphists regarded as results of partial alteration. Thus, in the case of associated crystals of andalusite and cyanite, Bischof does not hesitate to maintain the derivation of andalusite from the latter species by an elimination of quartz ; more than this, as the andalusite in question occurs in a granite-like rock, he suggests that itself is a product of the alteration of orthoclase. In like manner the mica, which in some cases coats tourmaline, and in others, fills hollow prisms of this mineral, is supposed to result from a subsequent alteration of crystallized tourmaline. So in the case of shells of leucite filled with feldspar, or of garnet enclosing epidote, or chlorite, or quartz, a similar transformation of the interior is supposed to have been mysteriously effected, while the external portion of the crystal remains intact. Again the aggregates of tinstone, quartz and orthoclase having the form of the latter, are, by Bischof and his school, looked upon as results of a partial alteration of previously formed orthoclase crystals. It needed only to extend this view to the crystals of calcite enclosing sand-grains, and regard these as the result of a partial alteration of the carbonate of lime. There is absolutely no proof that these hard crystalline substances can undergo the changes supposed, or can be absorbed and modified like the tissues of a living organism. It may, moreover, be confidently affirmed that the obvious facts of envelopment are adequate to explain all the cases of association upon which this hypothesis of pseudomorphism by alteration, has been based. Why the change should extend to some parts of a crystal and not to others, why in some cases the exterior of the crystal is altered, while in others the centre alone

---

* Report Geol. Survey of Canada, 1866, pp. 189, 213. Amer. Jour. Sci., III, i, 188.

is removed and replaced by a different material, are questions which the advocates of this fanciful hypothesis have not explained. As taught by Blum and Bischof, however, these views of the alteration of mineral species have not only been generally accepted but have formed the basis of the generally received theory of rock-metamorphism.

Protests against the views of this school have, however, not been wanting. Scheerer, in 1846, in his researches in Polymeric Isomorphism,* attempted to show that iolite and aspasiolite, a hydrous species which had been looked upon as resulting from its alteration, were isomorphous species crystallizing together, and, in like manner, that the association of olivine and serpentine in the same crystal, at Snarum in Norway, was a case of envelopment of two isomorphous species. In both of these instances he maintained the existence of isomorphous relations between silicates in which 3HO replaced MgO. He hence rejected the view of Gustaf Rose that these serpentine crystals were results of the alteration of olivine, and supported his own by reasons drawn from the conditions in which the crystals occur. In 1853 I took up this question and endeavored to show that these cases of isomorphism described by Scheerer, entered into a more general law of isomorphism pointed out by me among homologous compounds differing in their formulas by nM_2O_2 (M = hydrogen or a metal). I insisted, moreover, on its bearing upon the received views of the alteration of minerals, and remarked, "The generally admitted notions of pseudomorphism seem to have originated in a too exclusive plutonism, and require such varied hypotheses to explain the different cases, that we are led to seek for some more simple explanation and to find it, in many instances, in the association and crystallizing together of homologous and isomorphous species."† Subsequently, in 1860, I combated the view of Bischof, adopted by Dana, that "regional metamorphism is pseudomorphism on a grand scale," in the following terms : —

" The ingenious speculations of Bischof and others, on the possible alteration of mineral species by the action of various saline and alkaline solutions, may pass for what they are worth, although we are satisfied that by far the greater part of the so-called cases of pseudomorphism in silicates are purely imaginary, and, when

---

* Pogg. Annal., lxviii, 319.
† Pogg. Annal., lxviii, 319.

real, are but local and accidental phenomena. Bischof's notion of the pseudomorphism of silicates like feldspars and pyroxenes, presupposes the existence of crystalline rocks, whose generation this neptunist never attempts to explain, but takes his starting-point from a plutonic basis."

I then asserted that the problem to be solved in regional metamorphism is the conversion of sedimentary strata, "derived by chemical and mechanical agencies from the ocean-waters and pre-existing crystalline rocks into aggregations of crystalline silicates. These metamorphic rocks, once formed, are liable to alteration only by local and superficial agencies, and are not, like the tissues of a living organism, subject to incessant transformations, the pseudomorphism of Bischof." *

I had not, at that time, seen the essay by Delesse on Pseudomorphs already referred to, published in 1859, in which he maintained views similar to those set forth by me in 1853 and 1860, declaring that much of what had been regarded as pseudomorphism had no other basis than the observed associations of minerals, and that often "the so-called metamorphism finds its natural explanation in envelopment." These views he ably and ingeniously defended by a careful discussion of the whole range of facts belonging to the history of the subject.

My own expression of opinion on this question, in 1853, had been privately criticised, and I had been charged with a want of comprehension of the question. It was, therefore, with no small pleasure, that I not only saw my views so ably supported by Delesse, but read the language of Carl Friedrich Naumann, who in 1861 wrote to Delesse as follows, referring to his essay just noticed : —

"You have rendered a veritable service to science in restricting pseudomorphs to their true limits, and separating what had been erroneously united to them. As you have remarked, envelopments have, for the most part, nothing in common with pseudomorphs, and it is inconceivable that they have been united by so many mineralogists and geologists. It appears to me, moreover, that they commit an analogous error, when they regard gneisses, amphibolites, etc., as being, all of them, the results of metamorphic epigenesis, and not original rocks. It is precisely because pseudomorphism has been so often confounded with metamorphism that this error has found acceptance. I only admit a pseudomorph where there is some crystal the form of which has been preserved. There are very many metamorphic substances which are, in no sense of the word, pseudomorphs. Had the name of *crystalloid* been chosen, instead of pseudomorph, this confusion would certainly have never found its way into the science. I think, with you, that the envelopment of two minerals is most generally explained by a *contemporaneous* and *original* crystallization. Secondary envelopments, however, exist, and such may be called pseudomorphs or crystalloids, if they reproduce exactly the form of the crystal enveloped, whether this last still remains, or has entirely disappeared." *

It is unnecessary to remark that the view of Delesse and Naumann, viz. : that the so-called cases of pseudomorphism, on which the theory of metamorphism by alteration has been built, are, for the most part, examples of association and envelopment, and the result of a contemporaneous and original crystallization, — is identical with the view suggested by Scheerer, and generalized by myself long before, when, in 1853, I sought to explain the phenomena in question by "the association and crystallizing together of homologous and isomorphous species."

Later, in 1862, I wrote as follows : —

"Pseudomorphism, which is the change of one mineral species into another, by the introduction or the elimination of some element or elements, presupposes metamorphism (*i. e.*, metamorphic or crystalline rocks), since only definite mineral species can be the subjects of this process. To confound metamorphism with pseudomorphism, as Bischof, and others after him have done, is therefore an error. It may be farther remarked, that, although certain pseudomorphic changes may take place in some mineral species, in veins and near the surface, the alteration of great masses of silicated rocks by such a process is as yet an unproved hypothesis." †

Thus this unproved theory of pseudomorphism, as taught by Bischof, does not, even if admitted to its fullest extent, advance us a single step towards a solution of the problem of the origin of the various silicates, which, singly or intermingled, make up beds in the crystalline schists. Granting, for the sake of argument, that serpentine results from the alteration of olivine or labradorite, and steatite or chlorite from hornblende, the origin of

---

* Bull. Soc. Geol. de France, II, xviii, 678.

† Descriptive Catalogue, Crystalline Rocks of Canada, p. 80, London Exhibition, 1862; also, Dublin Quar. Journal, July 1863, and Amer. Jour. Sci., II, xxxvi, 218.

these anhydrous silicates, which are the subjects of the supposed change, is still unaccounted for. The explanation of this short-sightedness is not far to seek; as already remarked, Bischof, although a professed neptunist, starts from a plutonic basis. When the epigenic origin of serpentine and its related rocks was first taught, these were regarded as eruptive and unstratified, and it was easy to imagine intruded masses of dioritic and feldspathic rocks, which had become the subjects of alteration. As, however, the progress of careful investigation in the field has shown the stratified character of these serpentines, diallage-rocks, steatites, etc., and their intercalation among limestones, argillites, quartzites, gneisses, and mica-schists, and even among feldspathic and hornblendic strata, we are forced to reject, with Naumann, the notion of their epigenic derivation, and to regard them as original rocks.

This view brings us face to face with the problem of metamorphism as defined by me in 1860 * (*ante*, page 46). We must either admit that these crystalline schists were created as we find them, or suppose that they were once sands, clays, marls, etc.; in a word, sediments of chemical and mechanical origin, which by a subsequent process have been consolidated and crystallized. Whence, then, come these silicates of magnesia, lime, and iron, which are the sources of serpentine, hornblende, steatite, chlorite, etc.? This is the question which I proposed in that same year, when, after discussing the results of my examinations of the tertiary rocks near Paris, containing layers of a hydrous silicate of magnesia related to talc in composition, among unaltered limestones and clays, I remarked that it is evident "such silicates may be formed in basins at the earth's surface, by reactions between magnesian solutions and dissolved silica;" and, after some farther discussion, said "farther inquiries in this direction may show to what extent certain rocks composed of calcareous and magnesian silicates may be directly formed in the moist way."† Subsequently, in a paper on "The Origin of some Magnesian and Aluminous Rocks," printed in the Canadian Naturalist for June, 1860,‡ I repeated these considerations, referring to the well-known fact that silicates of lime, magnesia

---

*Amer. Jour. Sci., II, xxx, 135.
† Ibid., II, xxix, 284; also II, xl, 49.
‡ Ibid., II, xxxii, 286.

and iron-oxyd are deposited during the evaporation of natural waters, including those of alkaline springs and of the Ottawa River. Having described the mode of occurrence of the magnesian silicate, sepiolite, in the Paris basin, and the related quincite, containing some iron-oxyd and disseminated in limestone, I suggested that while steatite has been derived from a compound like sepiolite, the source of serpentine was to be sought in another silicate richer in magnesia; and, moreover, that chlorite, unless the result of a subsequent reaction between clay and carbonate of magnesia, was directly formed by a process analogous to that which (according to Scheerer) has, in recent times, caused the deposition from waters of neolite, a hydrous alumino-magnesian silicate approaching to chlorite in composition,[*] "the type of a reaction which formerly generated beds of chlorite in the same way as those of sepiolite or talc." Delesse, subsequently, in 1861, in his essay on Rock-Metamorphism insisted upon the sepiolites or so-called magnesian marls, as probably the source of steatite, and suggested the derivation of serpentine, chlorite, and other related minerals of the crystalline schists, from deposits approaching these marls in composition.[†] He recalled, also, the occurrence of chromic oxyd, a frequent accompaniment of these magnesian minerals, in the hydrated iron ores of the same geological horizon with the magnesian marls in France. Delesse did not, however, attempt to account for the origin of these deposits of magnesian marls, in explanation of which I afterwards verified Bischof's observations on the sparing solubility of silicate of magnesia, and showed that silicate of soda, or even artificial hydrated silicate of lime, when added to waters containing magnesian chlorid or sulphate, gives rise, by double decomposition, to a very insoluble magnesian silicate.[‡]

To explain the generation of silicates like labradorite, scapolite, garnet, and saussurite, I suggested that double aluminous silicates allied to the zeolites might have been formed, and subsequently rendered anhydrous. The production of zeolitic minerals observed by Daubrée at Plombières and Luxeuil by the action of a silicated alkaline water on the masonry of ancient Roman baths, was appealed to by way of illustration. It had there been shown

---

[*] Pogg. Annal., lxxi, 288.
[†] Etudes sur le Metamorphisme, quarto, pp. 91. Paris, 1861.
[‡] Amer. Jour. Sci., II, xl, 49.

by Daubrée that the elements of the zeolites had been derived in part from the waters, and in part from the mortar and even the clay of the bricks, which had been attacked, and had entered into combination with the soluble matters of the water to form chabazite. I, however, at the same time pointed out another source of silicated minerals, upon which I had insisted since 1857, viz. : the reaction between silicious or argillaceous matters and earthy carbonates in the presence of alkaline solutions. Numerous experiments showed that when solutions of an alkaline carbonate were heated with a mixture of silica and carbonate of magnesia, the alkaline silicate formed acted upon the latter, yielding a silicate of magnesia, and regenerating the alkaline carbonate ; which, without entering into permanent combination, was the medium through which the union of the silica and the magnesia was effected. In this way I endeavored to explain the alteration, in the vicinity of a great intrusive mass of dolerite, of a gray Silurian limestone, which contained, besides a little carbonate of magnesia and iron-oxyd, a portion of very silicious matter, consisting apparently of comminuted orthoclase and quartz. In place of this, there had been developed in the limestone, near its contact with the dolerite, an amorphous greenish basic silicate, which had seemingly resulted from the union of the silica and alumina with the iron-oxyd, the magnesia and a portion of lime. By the crystallization of the products thus generated it was conceived that minerals like hornblende, garnet and epidote might be developed in earthy sediments, and many cases of local alteration explained. Inasmuch as the reaction described required the intervention of alkaline solutions, rocks from which these were excluded would escape change, although the other conditions might not be wanting. The natural associations of minerals, moreover, led me to suggest that alkaline solutions might favor the crystallization of aluminous silicates, and thus convert mechanical sediments into gneisses and mica-schists. The ingenious experiments of Daubrée on the part which solutions of alkaline silicates, at elevated temperatures, may play in the formation of crystallized minerals, such as feldspar and pyroxene, were posterior to my early publications on the subject, and fully justified the importance which, early in 1857, I attributed to the intervention of alkaline silicates in the formation of crystalline silicated minerals. *

---

* Proc. Royal Soc., May 7, 1857. Amer. Jour. Sci., II, xxiii, 438, and xxv, 289 and 435.

While, however, there is good reason to believe that solutions of alkaline silicates or carbonates have been efficient agents in the crystallization and molecular re-arrangement of ancient sediments, and have also played an important part in that local alteration of sedimentary strata which is often observed in the vicinity of intrusive rocks, it is clear to me that the agency of these solutions is less universal than once supposed by Daubrée and myself, and will not account for the formation of various silicated rocks found among crystalline schists, such as serpentine, hornblende, steatite and chlorite. When I commenced the study of these crystalline strata I was led, in accordance with the almost universally received opinion of geologists, to regard them as resulting from a subsequent alteration of paleozoic sediments, which, according to different authorities, were of Cambrian, Silurian or Devonian age. Thus in the Appalachian region, as we have already seen, they have, on supposed stratigraphical evidence, been successively placed at the base, at the summit, and in the middle of the Lower Silurian or Champlain division of the New York system. A careful chemical examination among the unaltered paleozoic sediments, which in Canada were looked upon as the stratigraphical equivalents of the bands of magnesian silicates in these crystalline schists, showed me, however, no magnesian rocks except certain silicious and ferruginous dolomites. From a consideration of reactions which I had observed to take place in such admixtures in presence of heated alkaline solutions, and from the composition of the basic silicates which I had found to be formed in silicious limestones near their contact with eruptive rocks, I was led to suppose that similar actions, on a grand scale, might transform these silicious dolomites of the unaltered strata into crystalline magnesian silicates.

Farther researches, however, convinced me that this view was inapplicable to the crystalline schists of the Appalachians, since, apart from the geognostical considerations set forth in the previous part of this paper, I found that these same crystalline strata hold beds of quartzose dolomite and magnesian carbonate, associated in such intimate relations with beds of serpentine, diallage and steatite, as to forbid the notion that these silicates could have been generated by any transformations or chemical re-arrangement of mixtures like the accompanying beds of quartzose magnesian carbonates. Hence it was that already, in 1860, as shown above, I announced my conclusion that serpentine, chlorite and steatite had been derived from silicates like sepiolite, directly formed in waters at the earth's surface, and that the crystalline schists had resulted from the consolidation of previously formed sediments, partly chemical and partly mechanical in their origin. The latter being chiefly silico-aluminous, took, in part, the forms of gneiss and mica-schists, while from the more argillaceous strata, poorer in alkali, much of the aluminous silicate crystallized as andalusite, stauro-lite, cyanite and garnet. These views were reiterated in 1863,[*] and farther in 1864, in the following language, as regards the chemically-formed sediments: "steatite, serpentine, pyroxene, hornblende, and in many cases, garnet, epidote and other silicated minerals are formed by a crystallization and molecular re-arrange-ment of silicates generated by chemical processes in waters at the earth's surface."[†] Their alteration and crystallization was com-pared to that of the mechanically formed feldspathic, silicious and argillaceous sediments just mentioned.

The direct formation of the crystalline schists from an aqueous magma is a notion which belongs to an early period in geological theory. Delabeche in 1834[‡] conceived that they were thrown down as chemical deposits from the waters of the heated ocean, after its reaction on the crust of the cooling globe, and before the appearance of organic life. This view was revived by Daubrée in 1860. Having sought to explain the alteration of paleozoic strata of mechanical origin, by the action of heated waters, he proceeds to discuss the origin of the still more ancient crystalline schists. The first precipitated waters, according to him, acting on the anhy-drous silicates of the earth's crust, at a very elevated temperature, and at a great pressure, which he estimated at two hundred and fifty atmospheres, formed a magma, from which, as it cooled, were successively deposited the various strata of the crystalline schists.[§] This hypothesis, violating, as it does, all the notions which sound theory teaches with regard to the chemistry of a cooling globe, has, moreover, to encounter grave geognostical difficulties. The pre-Silurian crystalline rocks belong to two or more distinct sys-tems of different ages, succeeding each other in discordant strat-

* Geol. of Canada, pp. 577—581.
† Amer. Jour. Sci., II, xxxvii, 266, and xxxvlii, 183.
‡ Researches in Theoretical Geology, pp. 297-300.
§ Etudes et expériences synthétiques sur le Metamorphisme, pp. 119-121.

ification. The whole history of these rocks, moreover, shows that their various alternating strata were deposited, not as precipitates from a seething solution, but under conditions of sedimentation very like those of more recent times. In the oldest known of them, the Laurentian system, great limestone formations are interstratified with gneisses, quartzites and even with conglomerates. All analogy, moreover, leads us to conclude that even at this early period, life existed at the surface of the planet. Great accumulations of iron-oxyd, beds of metallic sulphids and of graphite, exist in these oldest strata, and we know of no other agency than that of organic matter, capable of generating these products.

Bischof had already arrived at the conclusion, which in the present state of our knowledge seems inevitable, that "all the carbon yet known to occur in a free state, can only be regarded as a product of the decomposition of carbonic acid, and as derived from the vegetable kingdom." He farther adds, "living plants decompose carbonic acid ; dead organic matters decompose sulphates, so that, like carbon, sulphur appears to owe its existence in a free state to the organic kingdom."[*] As a decomposition (deoxidation) of sulphates is necessary to the production of metallic sulphids, the presence of the latter, not less than that of free sulphur and free carbon, depends on organic bodies ; the part which these play in reducing and rendering soluble the peroxyd of iron, and in the production of iron ores is, moreover, well known. It was, therefore, that, after a careful study of these ancient rocks, I declared in May, 1858, that a great mass of evidence "points to the existence of organic life, even during the Laurentian or so-called azoic period."[†]

This prediction was soon verified in the discovery of the *Eozoön Canadense*, of Dawson, the organic character of which is now admitted by all zoologists and geologists of authority. But with this discovery, appeared another fact, which afforded a signal verification of my theory as to the origin and mode of deposition of serpentine and pyroxene. The microscopic and chemical researches of Dawson and myself showed that the calcareous skeleton of this foraminiferal organism was filled with the one or the other of these silicates in such a manner as to make it evident that

---

[*] Bischof, Lehrbuch, 1st. ed., II, 95. English ed., I, 252, 344.
[†] Amer. Jour. Science, II, xxv, 436.

they had replaced the sarcode of the animal, precisely as glauconite and similar silicates have, from the Silurian times to the present, filled and injected more recent foraminiferal skeletons. I recalled, in connection with this discovery the observations of
Ehrenberg, Mantell and Bailey, and the more recent ones of Pour-
tales, to the effect that glauconite or some similar substance occasionally fills the spines of Echini, the cavities of corals and mille-
pores, the canals in the shells of Balanus, and even forms casts
of the holes made by burrowing sponges (Clionia) and worms.
The significance of these facts was farther illustrated by showing
that the so-called glauconites differ considerably in composition,
some of them containing more or less alumina or magnesia, and
one from the tertiary limestones near Paris being, according to
Berthier, a true serpentine. *

These facts in the history of Eozoön, were first made known by
me in May, 1864, in the American Journal of Science, and subsequently more in detail, February, 1865, in a communication to the
Geological Society of London. †   They were speedily verified by
Dr. Gümbel, who was then engaged in the study of the ancient
crystalline schists of Bavaria, and soon recognized the existence,
in the limestones of the old Hercynian gneiss, of the characteristic *Eozoön Canadense*, injected with silicates in a manner precisely
similar to that observed by Dawson and myself. ‡   Later, in 1869,
Robert Hoffmann described the results of a minute chemical examination of the Eozoön from Raspenau, in Bohemia, confirming the
previous observations in Canada and Bavaria. He showed that
the calcareous shell of the Eozoön, examined by him, had been injected by a peculiar silicate, which may be described as related in
composition both to glauconite and to chlorite. The masses of
Eozoön he found to be enclosed and wrapped around by thin alternating layers of a green magnesian silicate allied to picrosmine,
and a brown non-magnesian mineral, which proved to be a hydrous silicate of alumina, ferrous oxyd and alkalies, related to
fahlunite, or more nearly to jollyte in composition. §

Still more recently, in the course of the present year, Dr. Dawson detected a mineral insoluble in acids, injecting the pores of

---

* Amer. Jour. Sci., II, xl, 360, Report Geol. Survey Canada, 1866, p. 231, and Quar.
Geol. Jour., XXI, 71.

† Amer. Jour. Sci., II, xxxvii, 431.   Quar. Geol. Jour., XXI, 67.

‡ Proc. Royal Bavar. Acad. for 1866, and Can. Naturalist, new series III, 81.

§ Jour. fur. Prakt. Chem., May, 1869, and Amer. Jour. Sci., III, i, 378.

crinoidal stems and plates in a paleozoic limestone from New Brunswick, which is made up of organic remains. This silicate which, in decalcified specimens, shows in a beautiful manner the intimate structure of these ancient crinoids, I have found by analysis to be a hydrous silicate of alumina and ferrous oxyd, with magnesia and alkalies, closely related to fahlunite and to jollyte.* The microscopic examinations of Dr. Dawson show that this silicate injected the pores of the crinoidal remains and some of the interstices of the associated shell-fragments, before the introduction of the calcite which cements the mass. I have since found a silicate almost identical with this, occurring under similar conditions in an Upper Silurian limestone said to be from Llangedoc in Wales.

Gümbel, meanwhile, in the essay on the Laurentian rocks of Bavaria, in 1866, already referred to, fully recognized the truth of the views which I had put forward, both with regard to mineralogy of Eozoön and to the origin of the crystalline schists. His results are still farther detailed in his *Geognost. Beschreibung des östbayerisches Grenzegebirges*, 1868, p. 833. Credner, moreover, as he tells us,† had already from his mineralogical and lithological studies, been led to admit my views as to the original formation of serpentine, pyroxene and similar silicates (which he cites from my paper of 1865, above referred to‡), when he found that Gümbel had arrived at similar conclusions. The views of the latter, as cited by Credner from the work just referred to, are in substance as follows:—The crystalline schists, with their interstratified layers, have all the characters of altered sedimentary deposits, and from their mode of occurrence cannot be of igneous origin, nor the result of epigenic action. The originally formed sediments are conceived to have been amorphous, and under moderate heat and pressure to have arranged themselves, and crystallized, generating various mineral species in their midst by a change, which, to distinguish it from metamorphism by an epigenic process, Gümbel happily designates *diagenesis*.

It is unnecessary to remark, that these views, the conclusions from the recent studies of Gümbel in Germany and Credner in North America, are identical with those put forth by me in 1860.

---

* Amer. Jour. Sci., III, i, 379.

† Hermann Credner; die Gleiderung der Eozoischen Formationsgruppe Nord Amerikas. Halle 1869. ‡ That in the Quar. Geol. Jour., XXI, 67.

At the early periods in which the materials of the ancient crystalline schists were accumulated, it cannot be doubted that the chemical processes which generated silicates were much more active than in more recent times. The heat of the earth's crust was probably then far greater than at present, while a high temperature prevailed at comparatively small depths, and thermal waters abounded. A denser atmosphere, charged with carbonic acid gas, must also have contributed to maintain, at the earth's surface, a greater degree of heat, though one not incompatible with the existence of organic life. * These conditions must have favored many chemical processes, which, in later times, have nearly ceased to operate. Hence we find that subsequently to the eozoic times, silicated rocks of clearly marked chemical origin are comparatively rare. In the mechanical sediments of later periods certain crystalline minerals may be developed by a process of molecular re-arrangement — diagenesis. These are, in the feldspathic and aluminous sediments, orthoclase, muscovite, garnet, staurolite, cyanite and chiastolite, and in the more basic sediments, hornblendic minerals. It is possible that these latter and similar silicates may sometimes be generated by reactions between silica on the one hand, and carbonates and oxyds, on the other, as already pointed out in some cases of local alteration. Such a case may apply to more or less hornblendic gneisses, for example, but no sediments, not of direct chemical origin, are pure enough to have given rise to the great beds of serpentine, pyroxene, steatite, labradorite, etc., which abound in the ancient crystalline schists. Thus, while the materials for producing, by diagenesis, the aluminous silicates just mentioned, are to be met with in the mud and clay-rocks of all ages, the chemically formed silicates, capable of crystallizing into pyroxene, talc, serpentine, etc., have only been formed under special conditions.

The same reasoning which led me to maintain the theory of an original formation of the mineral silicates of the crystalline schists, induced me to question the received notion of the epigenic origin of gypsums and magnesian limestones or dolomites. The interstratification of dolomites and pure limestones, and the enclosure of pebbles of the latter in a paste of crystalline dolomite, are of themselves sufficient to show that in these cases, at least, dolomites have not been formed by the alteration of pure limestones.

---

* Amer. Jour. Sci., II, xxxvi, 396.

The first results of a very long series of experiments and inquiries into the history of gypsum, were published by me in 1859, and farther researches, reiterating and confirming my previous conclusions, appeared in 1866.*   In these two papers, it will, I think, be found that the following facts in the history of dolomite are established, viz. : first, its origin in nature by direct sedimentation, and not by the alteration of non-magnesian limestones ; second, its artificial production by the direct union of carbonate of lime and hydrous carbonate of magnesia, at a gentle heat, in the presence of water.   As to the sources of the hydrous magnesian carbonate, I have endeavored to show that it is formed from the magnesian chlorid or sulphate of the sea or other saline waters in two ways :—first, by the action of the bicarbonate of soda found in many natural waters ; this, after converting all soluble lime-salts into insoluble carbonate, forms a comparatively soluble bicarbonate of magnesia, from which a hydrous carbonate slowly separates : second, by the action of bicarbonate of lime in solution, which, with sulphate of magnesia gives rise to gypsum ; this first crystallizes out, leaving behind a much more soluble bicarbonate of magnesia, which deposits the hydrous carbonate in its turn.   In this way, for the first time, in 1859, the origin of gypsums and their intimate relation with magnesian limestones were explained.

It was, moreover, shown that to the perfect operation of this reaction, an excess of carbonic acid in the solution, during the evaporation, was necessary to prevent the decomposing action of the hydrous mono-carbonate of magnesia upon the already formed gypsum.   Having found that a prolonged exposure to the air, by permitting the loss of carbonic acid, partially interfered with the process, I was led to repeat the experiment in a confined atmosphere, charged with carbonic acid, but rendered drying by the presence of a layer of dessicated chlorid of calcium.   As had been foreseen, the process under these conditions proceeded uninterruptedly, pure gypsum first crystallizing out from the liquid, and subsequently, the hydrous magnesian carbonate.†   This experiment is instructive as showing the results which must have attended this process in past ages, when the quantity of carbonic acid in the atmosphere greatly exceeded its present amount.

---

* Amer. Jour. Sci., II, xxxviii, 170, 365; xlii, 49.

† Proceedings Royal Institution, May 30, 1867, and Canadian Naturalist, new series, III, 231.

As regards the hypotheses put forward to explain the supposed dolomitization of previously-formed limestones by an epigenic process, I may remark that I repeated very many times, under varying conditions, the often cited experiment of Von Morlot, who claimed to have generated dolomite by the action of sulphate of magnesia on carbonate of lime, in the presence of water at a somewhat elevated temperature under pressure. I showed that what he regarded as dolomite was not such, but an admixture of carbonate of lime with anhydrous and sparingly soluble carbonate of magnesia; the conditions in which the carbonate of magnesia is liberated in this reaction, not being favorable to its union with the carbonate of lime to form the double salt which constitutes dolomite. The experiment of Marignac, who thought to form dolomite by substituting a solution of chlorid of magnesium for the sulphate, I found to yield similar results, the greater part of the magnesian carbonate produced passing at once into the insoluble condition, without combining with the excess of carbonate of lime present. The process for the production of the double carbonate described by Ch. Deville, namely, the action of vapors of anhydrous magnesian chlorid on heated carbonate of lime, in accordance with Von Buch's strange theory of dolomitization, I have not thought necessary to submit to the test of experiment, since the conditions required are scarcely conceivable in nature. Multiplied geognostical observations show that the notion of the epigenic production of dolomite from limestone is untenable, although its resolution and deposition in veins, cavities, or pores in other rocks is a phenomenon of frequent occurrence.

The dolomites or magnesian limestones may be conveniently considered in two classes; first, those which are found with gypsums at various geological horizons; and second, the more abundant and widely distributed rocks of the same kind, which are not associated with deposits of gypsum. The production of the first class is dependent upon the decomposition of sulphate of magnesia by solutions of bicarbonate of lime, while those of the second class owe their origin to the decomposition of magnesian chlorid or sulphate by solutions of alkaline bicarbonates. In both cases, however, the bicarbonate of magnesia, which the carbonated waters generally contain, contributes a more or less important part to the generation of the magnesian sediments. The carbonated alkaline waters of deep-seated springs often contain, as is well known, besides the bicarbonates of soda, lime, and magnesia, compounds of iron, manganese, and many of the rarer metals in solution, and thus the metalliferous character of many of the dolomites of the second class is explained. The simultaneous occurrence of alkaline silicates in such mineral waters, would give rise, as already pointed out, to the production of insoluble silicates of magnesia, and thus the frequent association of such silicates with dolomites and magnesian carbonates in the crystalline schists is explained, as marking portions of one continuous process. The formation of these mineral waters depends upon the decomposition of feldspathic rocks by subterranean or subaërial processes, which were doubtless more active in former ages than in our own. The subsequent action upon magnesian waters of these bicarbonated solutions, whether alkaline or not, is dependent upon climatic conditions, since, in a region where the rainfall is abundant, such waters would find their way down the rivercourses to the open sea, where the excess of dissolved sulphate of lime would prevent the deposition of magnesian carbonate. It is in dry and desert regions, with limited lake-basins, that we must seek for the production of magnesian carbonates, and I have argued from these considerations that much of northeastern America, including the present basins of the Upper Mississippi and St. Lawrence, must, during long intervals in the paleozoic period, have had a climate of excessive dryness, and a surface marked by shallow enclosed basins, as is shown by the widely spread magnesian limestones, and the existence of gypsum and rock-salt at more than one geological horizon within that area.* The occurrence of serpentine and diallage at Syracuse, New York, offers a curious example of the local development of crystalline magnesian silicates in Upper Silurian dolomitic strata under conditions which are imperfectly known, and, in the present state of the locality, cannot be studied.†

Since the uncombined and hydrated magnesian mono-carbonate is at once decomposed by sulphate or chlorid of calcium, it follows that the whole of these lime-salts in a sea-basin must be converted into carbonates before the production of carbonated magnesian sediments can begin. The carbonate of lime, formed

---

* Geology of Southwestern Ontario, Amer. Jour. Sci., II, xlvi, 355.

† Geology of the 3d district of New York, 108–110, and Hunt on Ophiolites, Amer. Jour. Sci., II, xxvi, 236.

by the action of carbonates of magnesia and soda, remains at first dissolved as bicarbonate, and is only separated in a solid form, when in excess, or when required for the needs of living plants or animals; which are dependent for their supply of calcareous matter, on the bicarbonate of lime produced, in part by the process just described, and in part by the action of carbonic acid on insoluble lime-compounds of the earth's solid crust. So many limestones are made up of calcareous organic remains, that a notion exists among many writers on geology that all limestones are, in some way, of organic origin. At the bottom of this lies the idea of an analogy between the chemical relations of vegetable and animal life. As plants give rise to beds of coal, so animals are supposed to produce limestones. In fact, however, the synthetic process by which the growing plant, from the elements of water, carbonic acid and ammonia, generates hydrocarbonaceous and azotized matters, has no analogy with the assimilative process by which the growing animal appropriates alike these organic matters and the carbonate and phosphate of lime. Without the plant, the synthesis of the hydrocarbons would not take place, while independently of the existence of coral or mollusk, the carbonate of lime would still be generated by chemical reactions, and would accumulate in the waters until, these being saturated, its excess would be deposited as gypsum or rock-salt are deposited. Hence in such waters, where, from any causes, life is excluded, accumulations of pure carbonate of lime may be formed. In 1861 I called attention to the white marbles of Vermont, which occur intercalated among impure and fossiliferous beds, as apparently examples of such a process.*

It is by a fallacy similar to that which prevails as to the organic origin of limestones, that Daubeny and Murchison were led to appeal to the absence of phosphates from certain old strata as evidence of the absence of organic life at the time of their accumulation.† Phosphates, like silica and iron-oxyd, were doubtless constituents of the primitive earth's crust, and the production of apatite crystals in granitic veins, or in crystalline schists, is a process as independent of life as the formation of crystals of quartz or of hematite. Growing plants, it is true, take up from the soil or the waters dissolved phosphates, which pass into the skeletons

---

* Amer. Jour. Sci., II, xxxi, 402.
† Siluria, 4th ed., pp. 28 and 537.

of animals, a process which has been active from very remote periods. I showed in 1854 that the shells of Lingula and Orbicula, both those from the base of the paleozoic rocks and those of the present time, have (like Conularia and Serpulites) a chemical composition similar to the skeletons of vertebrate animals.* The relations of both carbonate and phosphate of lime to organized beings are similar to those of silica, which, like them, is held in watery solution, and by processes independent of life is deposited both in amorphous and crystalline forms, but in certain cases is appropriated by diatoms and sponges, and made to assume organized shapes. In a word, the assimilation of silica, like that of phosphate and carbonate of lime, is a purely secondary and accidental process, and where life is absent, all of these substances are deposited in mineral and inorganic forms.

I have thus endeavored to sketch, in a concise and rapid manner, the history of the earlier rock-formations of eastern North America, and of our progress in the knowledge of them; while I have, at the same time, dwelt upon some of the geognostical and chemical questions which their study suggests. With the record of the last thirty years before them, American geologists have cause for congratulation that their investigations have been so fruitful in great results. They see, however, at the same time, how much yet remains to be done in the study of the Appalachians and of our northeastern coast, before the history of these ancient rock-formations can be satisfactorily written. Meanwhile our adventurous students are directing their labors to the vast regions of western America, where the results which have already been obtained are of profound interest. The progress of these investigations will doubtless lead us to modify many of the views now accepted in science, and cannot fail greatly to enlarge the bound of geological knowledge.

---

* Amer. Jour. Sci., II, xvii, 236.